PRAISE FOR THE AUTHOR'S *TINY STRUCTURES*

"For ingenuity, thrift, and charm, Mr. Diedricksen's tiny structures are hard to beat."
—*New York Times*

PRAISE FOR THE PREVIOUS EDITION: *HUMBLE HOMES, SIMPLE SHACKS, COZY COTTAGES, RAMSHACKLE RETREATS, FUNKY FORTS*

"Using salvaged materials and power tools, Diedricksen creates cozy living spaces filled with artistic touches that you would never expect to see in an average trailer or treehouse. He'll create colorful patterns of blue and green glass circles on a wall by sawing wine bottles in half. He's been known to incorporate pickle jars into his designs. That round window that looks like a porthole on a ship? That used to be the window on a front-loading washing machine. It's *Mad Magazine* meets *This Old House.*" —Emily Sweeney, *The Boston Globe*

"If you like little houses as much as I do, then your friends probably think you're weird. The images and ideas in Deek's book are not only great porn for tiny house freaks like us, but they also serve to show your friends that there is someone out there whose fascination with the subject is even more perverse than yours." —Jay Shafer, Tiny Tumbleweed House Company

"*Humble Homes* . . . is a great book, I really enjoyed it, and it's packed with information much like the old *Whole Earth Catalog.*" —Lester Walker, author of *Tiny Houses* and *American Shelter*

"I was so wowed by this book. It's awesome. The information is practical and raw, and very unpretentious. The overall feel is so honest and authentic." —Gregory Paul Johnson, president of The National Small House Society and Resourcesforlife.com

"Think *Wayne's World* meets *This Old House.*" —Jon Kalish, NPR/National Public Radio DIY host

"Deek brings the element of fun to the tiny house movement. He's obviously having a great time with his life, and his easily accessible drawings will inspire many a reader—who would otherwise not have done so—to pick up a hammer and saw and build a treehouse, a fort, a shack, or a solar shower, and realize 'Hey, I can do that!'" —Lloyd Kahn, editor of Shelter Publications and author of *Shelter*, *HomeWork*, and *Builders of the Pacific Coast*

"Amazing. This is a must-have book for those who appreciate extreme (and wacky) simple solutions! Crammed full of what flows from Derek's wild imagination, *Humble Homes* . . . is part comic book, notebook, and alternative living encyclopedia all in one. Handwritten and illustrated, virtually every page is crammed full of black-and-white drawings and notes that describe a vast array of ideas and solutions for simple living." —Michael Janzen, Tinyhousedesign.com

"I would not be surprised if *Humble Homes* . . . were to become a mini classic of its own within the surprisingly well-organized, and recently more and more in the spotlight, world of 'tiny house' enthusiasts. The book reads like a demented Boy Scout's fantasy notebook of plans, and some of the designs are downright cuh-razy, but all in all, there's a lot here to get your own creative juices going. To sum up, the book is a lot of fun and whets my appetite for future episodes of Diedricksen's 'Tiny Yellow House' video series—I can't wait to see more of his off-the-wall designs come to life!" —Katherine Sharpe, *ReadyMade Magazine*

"Derek's book is a far cry from anything conventional. Deek aims to inspire with his ideas, ideas that may well earn his book a place in tiny house history. What he ends up doing is reconstructing the mind into accepting what constitutes shelter. Deek's book is important not so much because it is another entertaining zine produced by an overly creative young person, but because he is both fed by a movement and contributing a large chunk to it with his mind-bending, Houdini-like acts of radically small, home-built shelters." —Amanda Kovattana, *Déjà Vu Construction*

"*Humble Homes* . . . explores the possibilities of the small home—and is the single craziest, most unique book I have ever seen. It's like MacGyver and Bob Vila sat down to make a graphic novel but made a catalog for Building 19 instead . . . all illustrated in a wonderful and crazy style." —Gene Higgins, *The Worcester Pulse Magazine*

"*Humble Homes* . . . is a hoot, an education, and an inspiration, crammed with Diedricksen's designs. The book is obsessively illustrated with wacky cartoon or design drawings and is brimming with Yankee ingenuity, junkyard philosophy, and plenty of eye-rolling yucks. The funky energy of *Humble Homes* . . . reminds me a lot of Lloyd Kahn's 'Shelter' books, Malcolm Wells's solar architecture books, and other homemade home books from *The Whole Earth Catalog*." —Gareth Branwyn, *Make Magazine*

"Simply put, if you are the type of person who walks by a pile of curbside junk and thinks, 'I could make something out of that,' then *this is the book for you*! Derek Diedricksen has put together a comprehensive collection of rather compelling building alternatives, combining his own building experiences with his creative imagination and excellent pen-and-ink drawings. Even if you never pick up a hammer, *Humble Homes* . . . is a hugely fascinating journey from beginning to end." —David and Jeanie Stiles, authors of *Cabins: A Guide to Building Your Own Nature Retreat* and *Sheds: The-Do-It-Yourself Guide*

"Brilliant! I want to write a book like this. But I can't. Derek is my new best friend and he doesn't even know me. He will though, by God. He will . . ." —Peter Nelson, founder of The Treehouse Workshop Inc. and author of *Treehouses of the World*

"For me, each page of *Humble Homes* . . . is a mutant love child of my past year. *A Few Zines* meets *Tiny Houses*. I never imagined the two could coexist, but Deek did. For him, a hardcore playing Eagle Scout, the link between DIY construction and DIY publishing is seamless. And inspired. Without moaning about the ill winds blowing down both media and housing, Deek tackled his pages with 83 black pens; one after another they gave themselves to his one-man revolution." —Mimi Zeiger, author of *Micro-Green* and *Tiny Houses*

"Think David Stiles and Rube Goldberg on a caffeine bender. One of the more unique and entertaining micro-housing books out there." —Mark Schelary, indie zine author of *Who Forted?*

"Clearly displaying his upbringing on a diet of punk rock and hardcore zines, Diedricksen has offered up one of the more eclectic and DIY mini-house books on the market." —*The Wilmington Weekly Online*

"The book weirded me out a little. . . . Is it cool if I put up a fence?" —the author's neighbor

"With close to fifty low-cost, do-it-yourself fort, cabin, and shelter designs, Diedricksen's *Humble Homes* . . . is a veritable eye-tiring, carpal tunnel–inducing barrage of wildly original dimestore-pen sketches, with a wealth of designs bordering from the clever and ultra efficient to the outright insane." —*Fortpimp Magazine*

Tiny Houses, Simple Shacks, Cozy Cottages, Ramshackle Retreats, Funky Forts

And Whatever the Heck Else We Could Squeeze in Here

(Say that ten times fast!)*

Written, Concocted, Conjured, and Illustrated by

Derek "Deek" Diedricksen

Foreword by David and Jeanie Stiles

*Warning: Saying the above book title ten times fast may
cause dizziness, fatigue, shortness of breath, and hair loss.

LYONS PRESS

GUILFORD, CONNECTICUT

An imprint of The Rowman & Littlefield Publishing Group, Inc.
4501 Forbes Blvd., Ste. 200
Lanham, MD 20706
www.rowman.com

Distributed by NATIONAL BOOK NETWORK

British Library Cataloguing in Publication Information available

Library of Congress Control Number Available

ISBN 978-1-4930-4650-8 paperback)
ISBN 978-1-4930-4651-5 (e-book)

♾™ The paper used in this publication meets the minimum requirements of American National Standard for Information Sciences—Permanence of Paper for Printed Library Materials, ANSI/NISO Z39.48-1992.

The procedures in this text are intended for use only by persons with prior training in the field of building "tiny houses." In the checking and editing of these procedures, every effort has been made to identify potentially hazardous steps and to eliminate as much as possible the handling of potentially dangerous materials, and safety precautions have been inserted where appropriate. If performed with the materials and equipment specified, in careful accordance with the instructions and methods in this text, the author and publisher believe the procedures to be very useful tools. However, these procedures must be conducted at one's own risk. The author and publisher do not warrant or guarantee the safety of individuals using these procedures and specifically disclaim any and all liability arising directly or indirectly from the use or application of any information contained in this publication.

Deek's Disclaimer ('Cause, sadly, these days everyone's gotta have one)

In this age of "Gee, how was I to know that this steaming cup of McDonald's coffee could leave me looking like an unmasked Phantom of the Opera if I spilled it on myself? I'm suing!" I find it necessary to include the following disclaimer for what should stand as common sense anyway:

USE YOUR BRAIN.
DON'T BE AN IDIOT.

READ.
BUILD.
DO SOMETHING
USE YOUR BRAIN!
(FEW DO.)

A good many of these structures are stylistic recommendations and nothing more. Since the plans are loosely (or not even) laid out, when and *if* you attempt to build any of these, a good deal of your own freethinking, problem-solving, and personal skill level will come into play, so build *only* what you think you are capable of. If you build any of these cabins and they collapse on you, or you hacksaw your thumb right off, I'm in no way responsible. It's your arse! (Try to keep that too. It comes in handy when sitting.) And yeah, some of these structures border on the ridiculous, so use extra care when treading in that territory.

It goes without saying that this entire book and all its art may not be copied, distributed, or sold without explicit permission from the author and illustrator (aka me) and Lyons Press. If you choose to ignore this simple, lawful request, I will fly out to wherever you are and mess you up. I'm a ninjitsu master in the art of West Coast spatula fighting, so don't push me!

CONTENTS

FOREWORD

It was early in 2010 when we received a fledgling copy of *Humble Homes, Simple Shacks*—a comb-bound book then in its original, self-released "basement form." Well, what at first seemed like some kind of wacky comic book soon took on another form as we began leafing through its enormous array of details and design concepts. We found ourselves reading every page, blurb, micro-sketch, and side note of the book.

Simply put, if you are the type of person who walks by a pile of curbside junk and thinks "I could make something out of that," then this is the book for you! Derek Diedricksen has put together a comprehensive collection of rather compelling building alternatives, combining his own building experience with his creative imagination and excellent pen and ink drawings. Derek is the modern-day equivalent of the beat generation builders who, during the 1960s, looked for new and inexpensive ways to build their own shelters. Encouraged by books and magazines like Ken Kern's *The Owner Built Homestead*, *The Whole Earth Catalog*, *Shelter* magazine, and *Mother Earth News*, these disenfranchised young people with little or no building experience began constructing and living in geodesic domes, yurts, teepees, and log cabins. Diedricksen's work continues very much in this vein.

Nowadays, with unemployment at its highest level and foreclosures on farms and properties increasing, the time seems right for ambitious young people to think about building their own homes, if they can afford to buy a small piece of property. Sound ridiculous? Many people, including ourselves, have done just that by using their vacations and weekends and a small part of their salaries to build their own getaway cabins. In order to do this you need some type of temporary shelter to sleep in and store your tools, and this is where Derek's book can be of great assistance. And for those who aim to dwell permanently in the extremes of micro-architecture, look no further. When we bought our land 50 miles north of New York City, we built a small temporary 12-by-12-foot cabin in the woods, figuring this would be good enough while we were deciding on the house site and what it would look like. We used the cabin on weekends to go fishing, swimming, hiking, and cross-country skiing. As it turned out, we never got around to building the real house because we had everything we needed already! Perhaps this book will help you realize the same thing and will inspire you to build your own cabin. Even if you never pick up a hammer, *Humble Homes* is a hugely fascinating journey from beginning to end.

—David and Jeanie Stiles,
authors of *Cabins: A Guide to Building Your Own Nature Retreat* and *Sheds: The-Do-It-Yourself Guide*

PREFACE: FROM HERE TO THERE:
A-RAGS-TO-FANCIER-RAGS STORY

Humble Homes, Simple Shacks, Cozy Cottages, Ramshackle Retreats, Funky Forts, and Whatever the Heck Else We Could Squeeze in Here was initially, and hesitantly, released in late 2009 under the publishing moniker Tiny Yellow House Press. In reality, TYH Press was run by one man, Derek Diedricksen, on his ten-dollar tag-sale comb-binding machine from the depths of his musty basement just outside of Boston. The books were first sold by word of mouth, then on the tiny-house blogosphere, and finally on Diedricksen's own website, Relaxshacks .com. The entire undertaking began with trash in the form of recycled cans, as you'll later read.

To Derek's complete surprise, a book he almost didn't release rapidly gained an indie following and escalated, all within a year, to what some might call "Instant Cult-dom." He had no formal background in architecture or license in carpentry, and no one was more surprised by the book's success than the author himself.

Humble Homes . . . soon gained attention from the likes of National Public Radio, *Make Magazine*, *ReadyMade Magazine*, CBS Radio, and local television stations, which propelled the book to a whole new level. Meanwhile, a web/TV show dubbed *Tiny Yellow House*, based on the designs in this underground book, was launched, and the first slapped-together episode of the show received more than eight thousand views in its first week alone. The rest is tiny house history . . .

After offers from three publishing companies, Derek ultimately handed over the reins to Lyons Press to produce his book. While trying to juggle his roles as host/director/producer of *Tiny Yellow House TV*, head contractor of his own micro-carpentry business, drummer for two bands, key writer for three blogs, and father of two small kids, he found himself less and less able to keep up with online book orders. (Did we mention that each and every book was hand-assembled?)

The book you now hold in your hands, in the words of the author, is a "greatly expanded, more thoughtfully laid-out, color-enriched, reworked version of the book that took hours upon hours of painstaking work—and lots of coffee—to complete." One can only imagine . . .

May you enjoy and find inspiration within its bizarre, unique contents.

—"B. Grizz Lee"
(an off-the-radar tiny house enthusiast/
dweller/blogger somewhere out in Idaho)

INTRODUCTION: TO THE READER ABOUT TO EMBARK UPON A JOURNEY THROUGH A SLEW OF DOODLES (AKA MY "GET-RICH-SLOW SCHEME")

First off, and to put it bluntly, I must tell you that this book does not include spoon-fed building diagrams and materials lists for all the projects and ideas within, as they are just that: ideas that I hope will inspire you. Better yet, you might even find a way to improve upon these recommendations, or take them on some architectural tangent. On the other hand, to those of you who are neither creative nor open-minded enough to fit any of these loose designs to your own needs and measurements, maybe you should look into hiring someone to blow your nose for you—another subject more appropriate for my next book, *The Do-It-Yourself Guide to Doing Nothing Yourself* (I know at least three dozen people who'd preorder copies).

What I'm getting at is that in every other review for housing "idea" books, there's always some obsessive-compulsive, automatonical knucklehead who complains that the drawings and measurements given are not ultra-specific enough.

For example:

Review by Blogaholic43: Diedricksen's book, while rather juvenile and brash, was unsatisfactory due to the fact that the plans displayed failed to advise which lockset and peephole combination and brand of front door to use. And what about opening said doors? How am I to do that without explicit instructions? Help!

Still trapped outside my own home,
Blooblestein Finklebrats III

Again, you're holding the wrong book if you need details as concise as those. This is not a blueprint book. Sorry. On the positive side, though, I have come close to digitally crippling myself in order to provide a decent and honest—albeit unschooled and possibly unconventional—book to add to your tiny-home-junkie collection. As I've fatigued my eyes and blistered my fingers, taking hours upon hours to piece these sketches together (never mind the hundreds of hours spent drawing the illustrations!), I hope you realize that if I were in this for the money, I wouldn't have tackled this book at all. Heck, if I were in this for the money, I would have stopped typing out this longwinded introduction long ago.

To all the naysayers, complainers, anonymous e-mail and Internet tough guys, and unstable, negative creeps out there (most of them my relatives), who shall never be pleased or satisfied with enough tedious details, might I suggest you:

1. Head-butt an anvil at least five or six times to see who'll win. If you lose, please try and try again, as no one likes a quitter!
2. See if your fingers can stop a moving Cuisinart blade.
3. Use Krazy Glue as a lip gloss substitute.
4. Go huff a toilet vent-stack.
5. Take a piranha bubble bath.
6. Make out with a cobra.
7. Skinny-dip in an active volcano.

And, since this *is* a new and expanded version of the book:

8. Buy yourself a copy of *Gigli*, and actually (shudder) watch it.
9. Binge on Diet Coke and Mentos while riding a pogo stick.
10. Catch a donkey kick with your groin.
11. Make fun of Mike Tyson's lisp—to his face.

But in all seriousness, and before any of my sarcastic rants scare you away (and before Mike Tyson sees this and buries me), I do appreciate your giving this grassroots, semi-disorganized ream of cartoonish sketches a shot, and I hope that you find something worthwhile within these pages. Ultimately, if this book reaches and inspires even one kid-at-heart to build any of the included outlines, I'll consider this book a success (and celebrate with a feast of ramen noodles bought with the residuals and royalties).

So, back off if you don't like this book, and keep in mind that instead of loafing around in stained, tie-dyed Smurf sweatpants, cultivating a Unabomber beard, and sacrificing my retinas to excessively violent video games (the American Way!), I took the time to do something constructive—and it wasn't for the ducats (or Benjamins) either.

Although, if I may be honest, I have had my eye on a certain Lamborghini, covered in gold leaf and rhinestones, with a puffy-paint "D" on the hood (and now that I've signed a publishing deal, that Matchbox toy is mine!).

THE WHY

"Why the heck did you decide to write a book on *this* subject?"

This is the question I'm most frequently asked, aside from, "What cheap, tainted, street drugs were you on when you decided to go into the field of micro-architecture and salvage construction?"

Well, the answer is that I've been fascinated with compact living ever since my father gave me a

A MICRO, RED-GLASS CANDLE-LANTERN THAT I FOUND NEW FOR $1.00. BIZARRELY, SOME OF MY MICRO-CABIN IDEAS START WITH ACCENT ITEMS LIKE THIS.

THE NUMBER ONE QUESTION IS USUALLY "THE NUMBER TWO" QUESTION...

WE'LL GET TO THAT LATER.

hardcover copy of Lester Walker's *Tiny, Tiny Houses* when I was no more than ten. I also had some experience with living in tight spaces, as I grew up in a very modest ranch house in Madison, Connecticut, where I shared (the horror! egads!) a 12-by-12-foot room with my brother until ninth grade. This was in a well-to-do town, by the way, where every other kid I knew had a bedroom for each season. Splitting living quarters was pretty darn easy and unproblematic—heck, enjoyable even—although retiring to a tiny, shared bedroom after family burrito night did have its downside.

I also have my grandfather, Alvin "King of the Salvaged Nails" Clement Diedricksen (1921–2007) to thank. One of the thriftiest men I've known, he was the original king of the junk-loaded "Super Garage," as he called it. We're still unearthing unfathomable odds and ends from its depths—heck, for all we know, Jimmy Hoffa could be in there. (Go Browns!

"Tweet Tweet!") He also introduced me to *Mad* magazine, an early influence that I'm sure has played a large role in this book's existence.

The love of tiny homes has carried over to my brother, Dustin, as well. He and his wife (and 618 pets, give or take) live in a very modest 800-square-foot home. You'll see a few of Dustin's tiny house tips crammed into this book too!

Anyway, having come from such a background, I must admit that I find it a little troublesome to talk to people who feel they *need* to live in a 7,000-square-foot McMansion on a street devoid of trees, all the while pissin' away oil, propane, and natural gas just to keep their collection of dusty golf trophies and *Sex and the City* DVDs warm. It goes without saying that this book isn't for you. However, I'm sure you will never pick up this book or read these words, so I think I'm safe from the vengeful path of your Lexus.

Vehicular homicide aside, while there are numerous small home, cabin, and housing books out there right now, from indie and mega-corporate publishers alike (many of which I recommend in the Book Nook section of this tome), I'm a total and complete small-home book junkie, and I've read just about all of the decent ones I've been able to get my grubby mitts on!

Seriously.

The problem with such an addiction is that after a short while you've run out of new options and you're left craving more. Even worse, there seems to be a glut of authors and publishers groping for big sales in the new hip and ultra-sellable modest- and sustainable-living fields. A good many of these books satisfy the subject in title alone. I guarantee that almost half the books on "small houses" in any bookstore won't feature many homes under *2,000* square feet. Good lord, 2,000 square feet? The denizens of such "hovels" surely deserve a pat on the back for toughing it out! While these books carry titles claiming true insight into "micro living," some more truthful titles would be:

I've Finally Caught Up to the Joneses, and Boy Will My Grandkids Be Pissed When They're Still Paying for It

You Too Can Live Like an Absolute Moron for Only $400,000 Down!

Yeah, You're "Green," but Only Because You Ate Some Steak Tartare That Didn't Agree with You

Fourth Vacation Homes, and the Wasteful, Vulpine, Brandy-Snifter-Toting Attorneys Who Never Use Them

And . . . *Bulldozing That Perfectly Charming Home to Make Way for a Newer, Bigger, Far-Uglier One*

But let me get back on track here. This book aims to provide one more option for those truly interested in tiny homes and dwellings, and to showcase a few others who've built modest digs of their own. Yes, I know, you'll find some of these designs utterly ridiculous or close to unbuildable, but I had one heck of a time brainstorming all of them. May the net proceeds of this book cover the carpal tunnel surgery I will need after all these damn dime-store-pen sketches.

I hope this book sparks something in you—perhaps something that could lead to another tiny-house book for the shelves of my own library. Now get off your butt (or your head, if you're perusing this text in some weird upside-down yoga posture)! It all invites the question: "How much space do you really, truly, need to live well?"

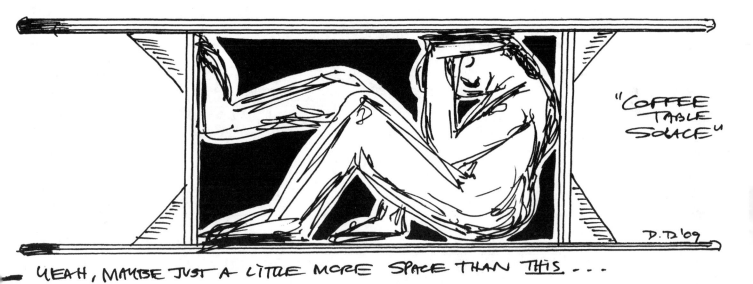

"COFFEE TABLE SOLACE"

D.D '09

YEAH, MAYBE JUST A LITTLE MORE SPACE THAN THIS . . .

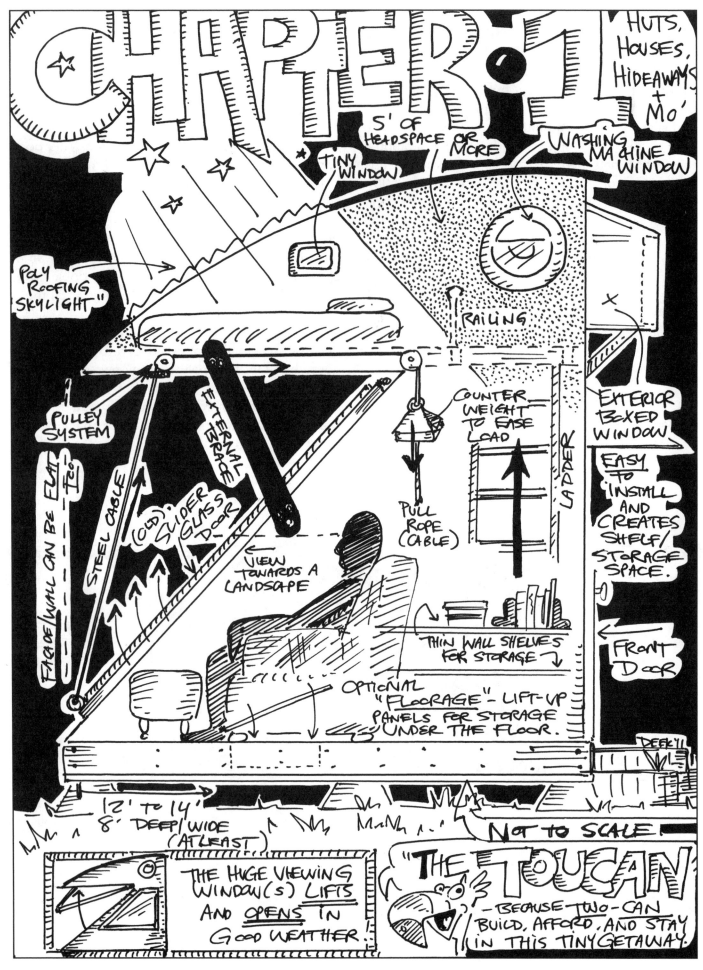

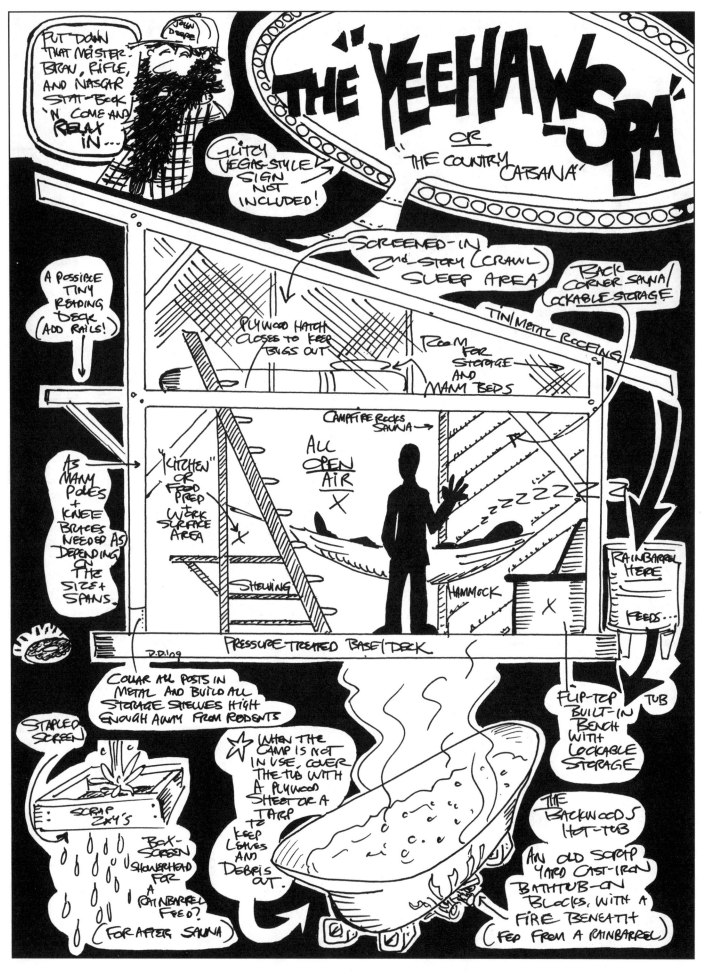

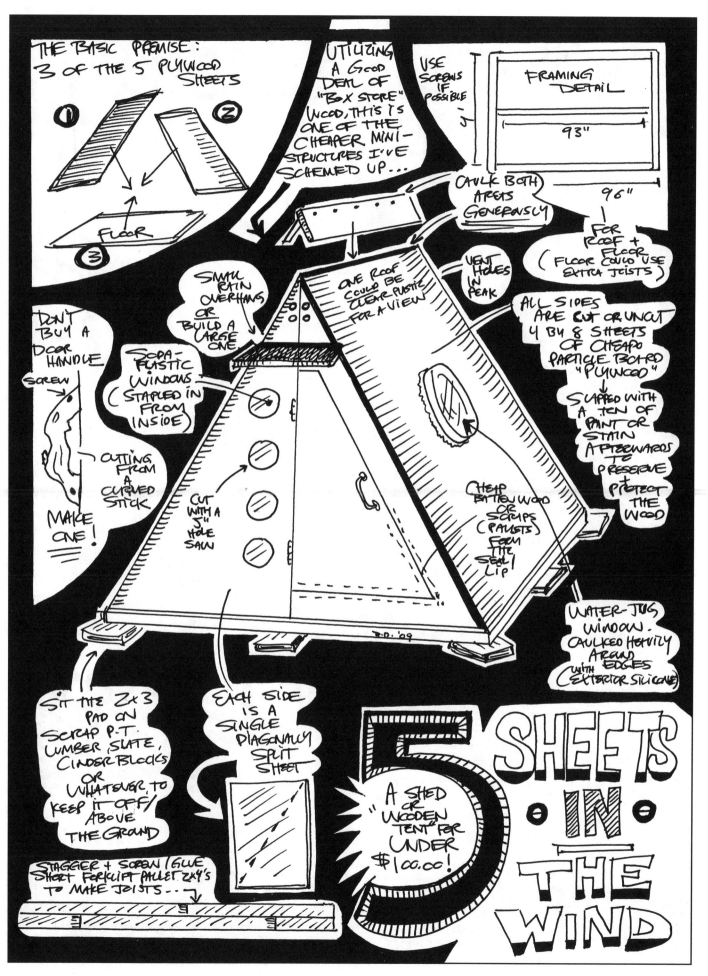

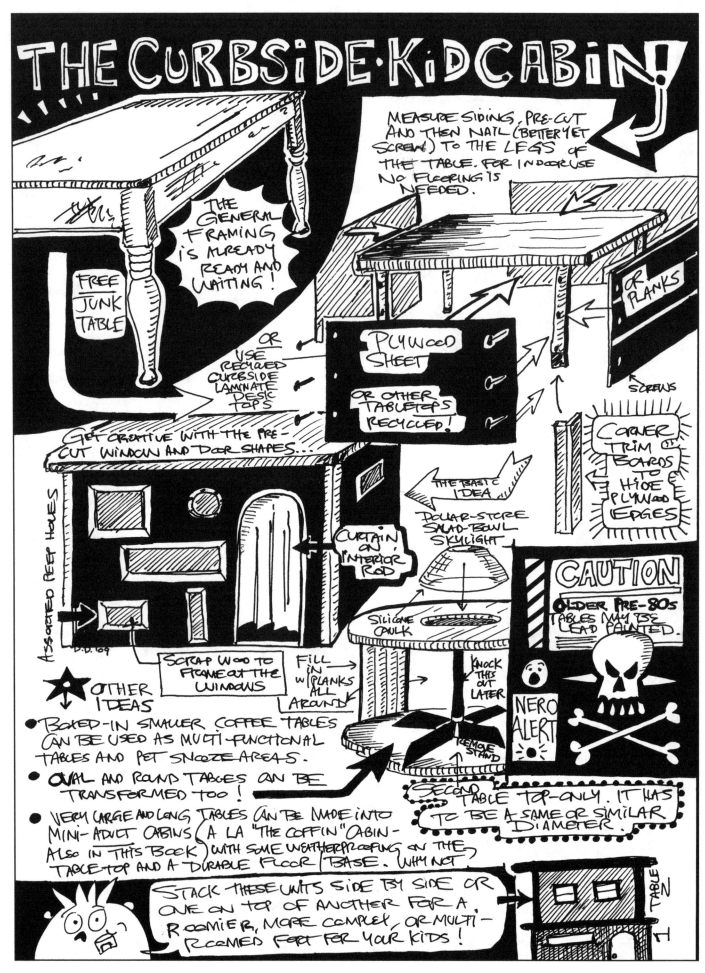

THE CURBSIDE·KID CABIN!

MEASURE SIDINGS, PRE-CUT AND THEN NAIL (BETTER YET SCREW) TO THE LEGS OF THE TABLE. FOR INDOOR USE NO FLOORING IS NEEDED.

THE GENERAL FRAMING IS ALREADY READY AND WAITING!

FREE JUNK TABLE

OR USE RECYCLED CURBSIDE LAMINATE DESK TOPS

OR PLANKS

PLYWOOD SHEET

OR OTHER TABLETOPS RECYCLED!

SCREWS

CORNER TRIM BOARDS TO HIDE PLYWOOD EDGES

GET CREATIVE WITH THE PRE-CUT WINDOW AND DOOR SHAPES...

ASSORTED PEEP HOLES

CURTAIN ON INTERIOR ROD

THE BASIC IDEA

DOLLAR-STORE SALAD-BOWL SKYLIGHT

SILICONE CAULK

D.D.'09

SCRAP WOOD TO FRAME OUT THE WINDOWS

FILL IN W/ PLANKS ALL AROUND

KNOCK THIS OUT LATER

REMOVE STAND

CAUTION

OLDER PRE-80s TABLES MAY BE LEAD PAINTED.

NERD ALERT

★ OTHER IDEAS

- BOXED-IN SMALLER COFFEE TABLES CAN BE USED AS MULTI-FUNCTIONAL TABLES AND PET SNOOZE AREAS.

- OVAL AND ROUND TABLES CAN BE TRANSFORMED TOO!

- VERY LARGE AND LONG TABLES CAN BE MADE INTO MINI-ADULT CABINS (A LA "THE COFFIN" CABIN - ALSO IN THIS BOOK) WITH SOME WEATHERPROOFING ON THE TABLE-TOP AND A DURABLE FLOOR/BASE. WHY NOT?

SECOND TABLE TOP-ONLY. IT HAS TO BE A SAME OR SIMILAR DIAMETER.

STACK THESE UNITS SIDE BY SIDE OR ONE ON TOP OF ANOTHER FOR A ROOMIER, MORE COMPLEX, OR MULTI-ROOMED FORT FOR YOUR KIDS!

TABLE 2

TABLE 1

18

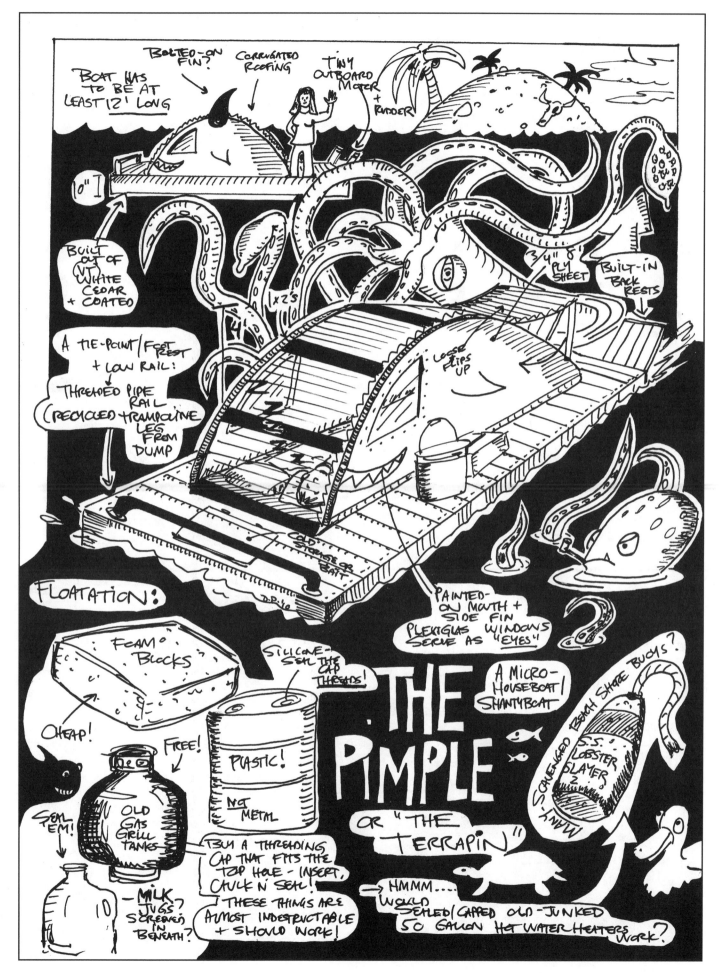

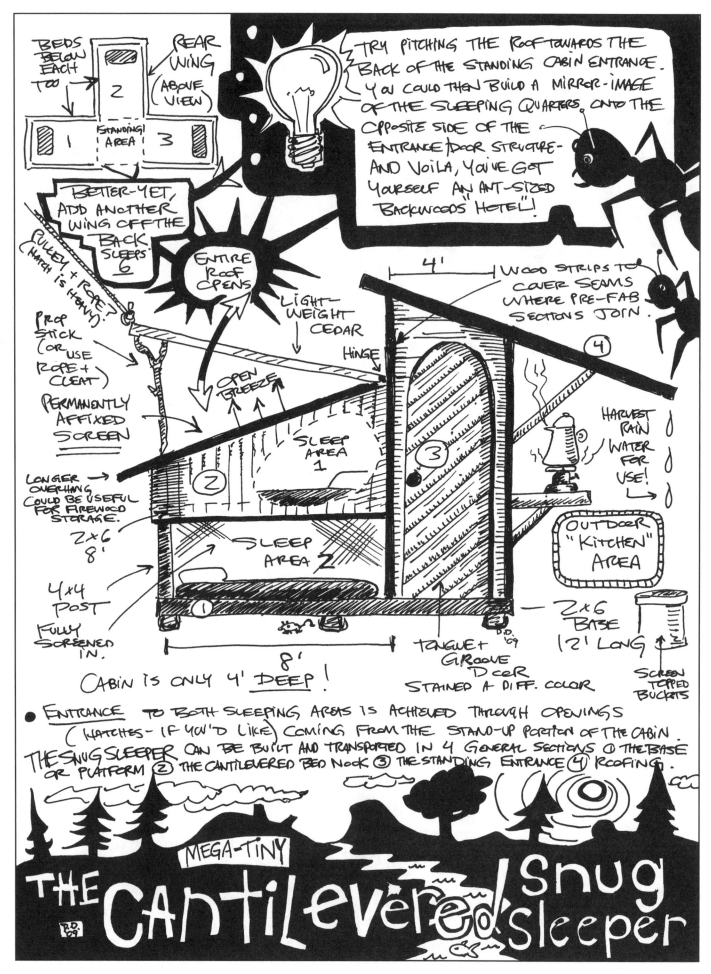

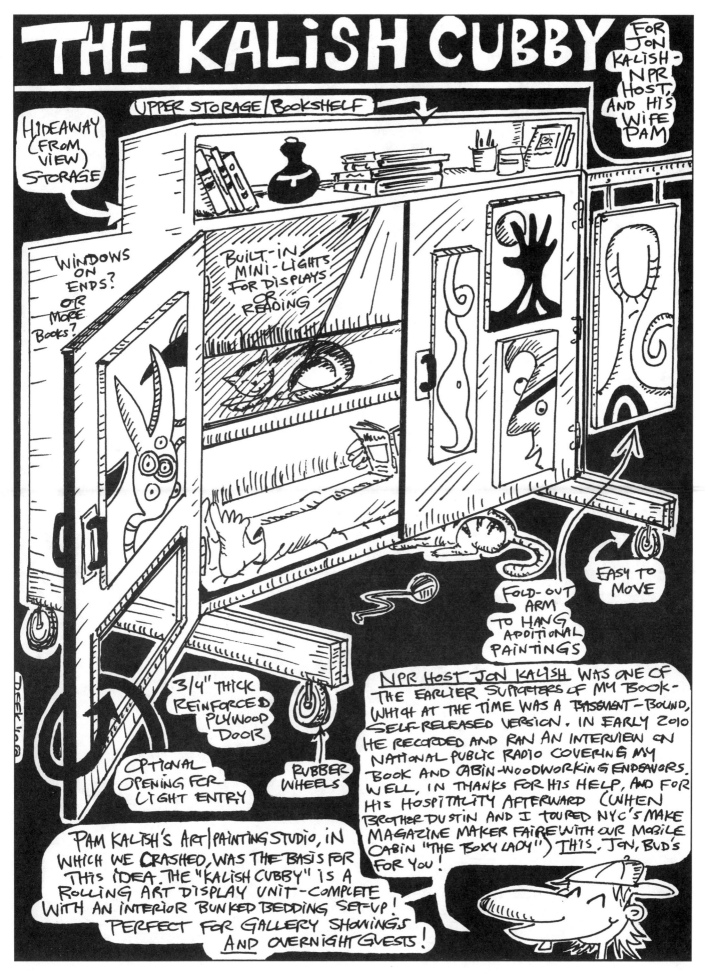

THE OWL'S NEST

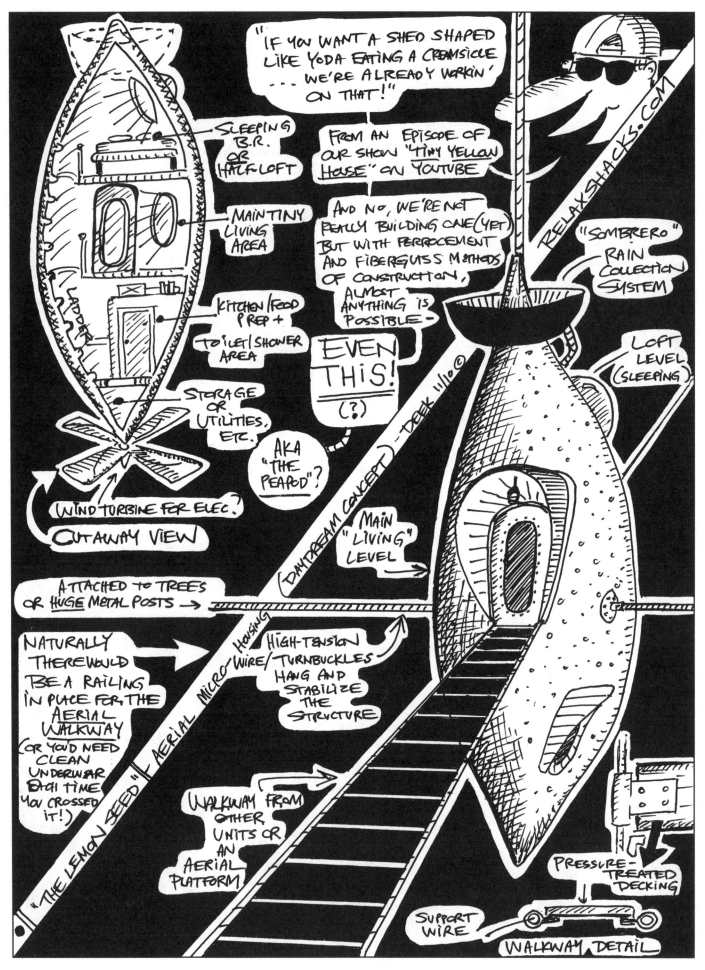

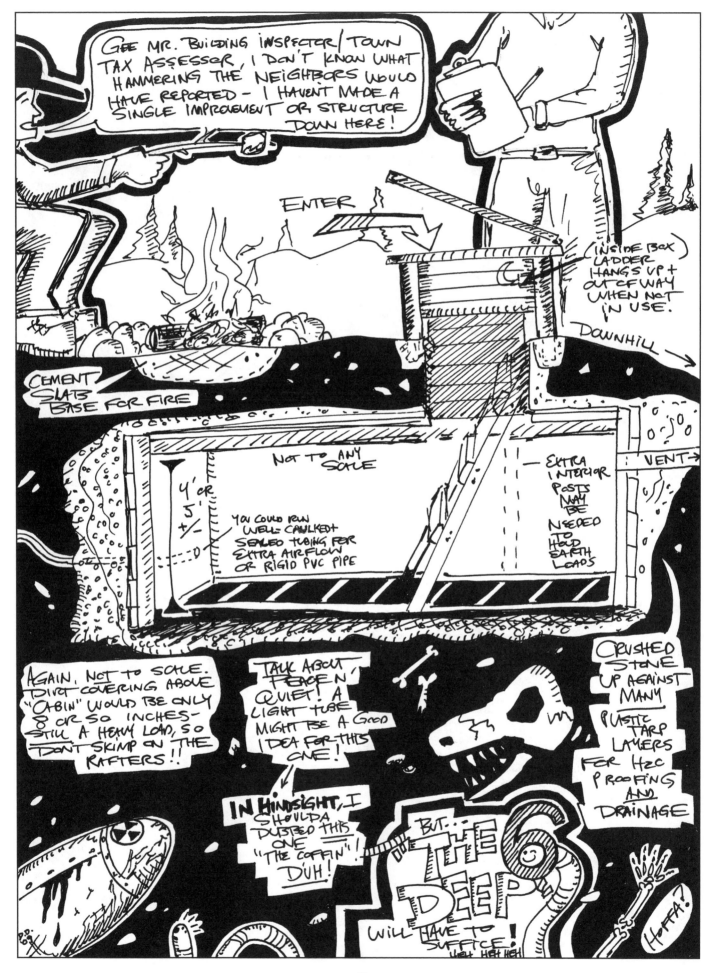

THIS ONE'S SURE TO GET THE EYES OF NEIGHBORS/FRIENDS/ VISITORS/BUILDING AUTHORITIES ROLLING, AND ITS A LEFT-FIELD, UNCONVENTIONAL IDEA, BUT C'MON NOW, WHAT A UNIQUE, FUN, AND RELAXING PLACE A SKY-HIGH READING NEST WOULD BE! AND NAPPING FORTY FEET OFF THE GROUND AMIDST THE SOFT RUSSLIN' OF BLOWING LEAVES, AND CHIRPING BIRDS... I COULD CERTAINLY GO FOR THAT!

(RUSTLING)
IN REDNECK SPEAK

EYE-SCREW OR TIED

ROPE RAIL

2×12 WALKWAY (ACCESS) FROM ANOTHER TREE AND THE INITIAL LADDERWAY UP.

DRINK HOLDER. OLD TIN CAN NAILED TO TREE WORKS TOO!

HOME WORK -KAHN

NOW THIS IS WHAT I CALL "HIGHER LEARNING"!

GAPS, HOLES + NON-HORIZONTAL BOARDS ARE HALF THE REDNECK CHARM, I MEAN C'MON!

D.D. '09

BE IT THROUGH THE (THRIFTY/CHEAPO) DEPRESSION-ERA WAYS I INHERITED THROUGH ALVIN DIEDRICKSEN ("KING OF THE SALVAGED NAILS") OR THROUGH OTHER OUTSIDE FORCES OVER THE YEARS, I'M A FULL-OUT LOVER/SUCKER FOR ANY STRUCTURE MADE OUT OF "PURE SCRAP 'N CRAP", AS THIS ONE IS. ALL THESE TOO-SMALL-TO-USE 2×4 AND SCRAP WOOD ENDS.. ...YUP, HERE'S WHERE THEY'RE GIVEN SECOND LIFE. AND AS I DO KEEP COFFEE CANS FULL OF "SALVAGEABLE NAILS", YOU MIGHT AS WELL PUT THEM TO USE IN A PROJECT LIKE THIS ONE IF YOU DO THE SAME.

THICK ROPE FOR EASIER/QUICKER EXITING

NESTING OF A DIFFERENT SORT

★ TURN THAT BURN PILE WOOD INTO A "BUILD PILE"!

FORKLIFT PALLET WOOD, LEFTOVER FRAMING ENDS, PICKETS FROM AN OLD FENCE, OR THE REMAINS OF A DEMO'D DECK - ANYTHING IS FAIR GAME HERE!

PAINT THE EXTERIOR WITH A QUICK CAMOUFLAGE-EARTH-COLORED COAT IF YOU WANNA GET "FANCY".

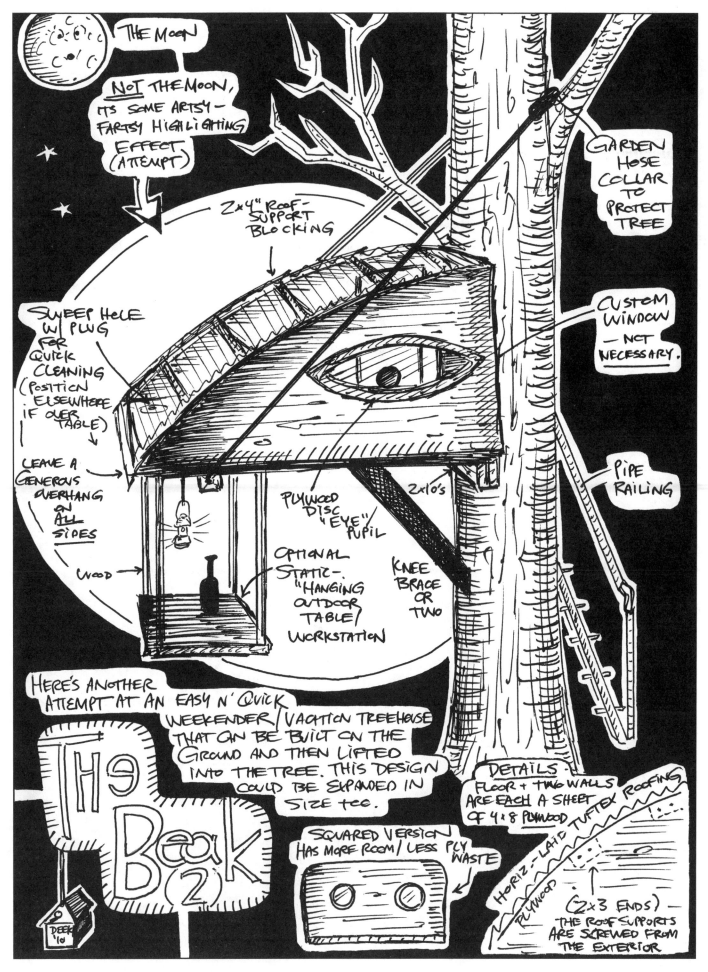

"SUPER-SIMPLE N' SMALL SHED-ROOFED SHACK"
A MEGA-CHEAP VACATION GETAWAY!

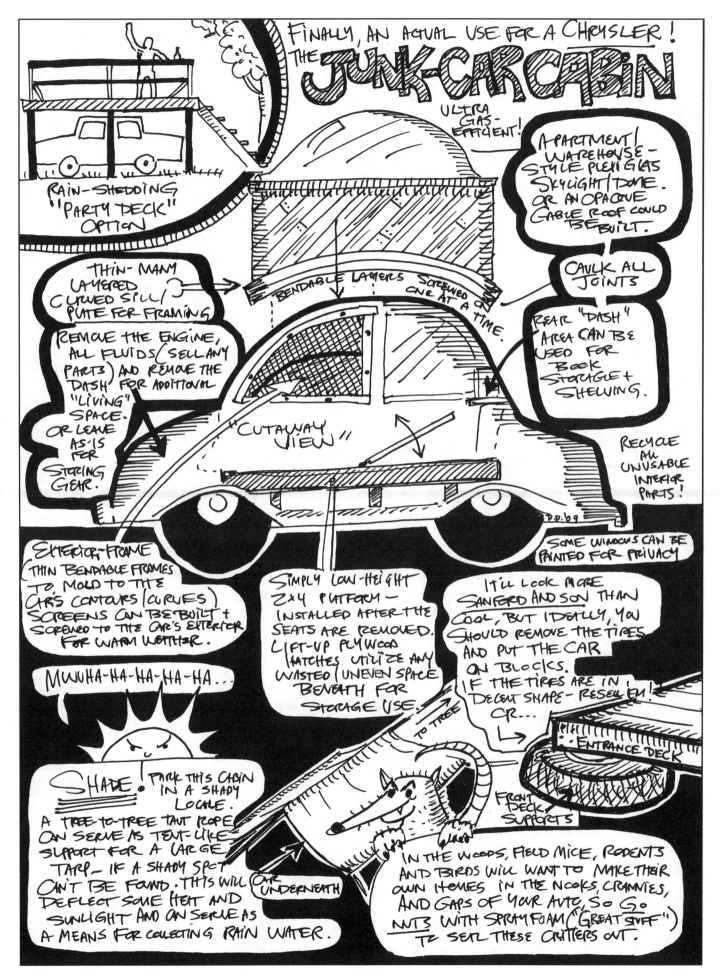

THE "A-FRAME"

... AND "THE HALF-A"

DRIP OVERHANG

ALL-WINDOWED FRONT

EXTERIOR PORCH

LOFT

INEXPENSIVE T-111 SIDING PERHAPS

LESS INTERIOR SPACE OVERALL, BUT YOU'RE GIVEN VERTICAL SPACE FOR WINDOWS AND NATURAL LIGHTING AS A TRADE-OFF - NEVERMIND THE INTERESTING LOOK.

INTERIOR

STORAGE

EASY DOWN-THE-ROAD SHED ROOF ADDITION

HUMANURE TOILET?

YOU COULD EVEN ALTER THE IDEA AND BUILD A 2nd STORY DECK - ACCESSED BY THE LOFT.

16' STOCK LUMBER LENGTHS

BEDS ALL DOUBLE AS COUCH-STYLE SEATING

10' OR 12'

2'

4'

THE OLE CLASSIC:
EASY TO BUILD, EASY TO PRE-BUILD, AND A GREAT OPTION FOR AREAS WITH HEAVY SNOWFALL (THE STEEP ROOF READILY SHEDS DANGEROUS SNOWLOADS.

FRONT PORCH

BED 1

BED 2

X
(2nd FLOOR/LOFT)

LADDER

BED 3

WOODSTOVE

SHELVING

THE POTENTIAL DIEDRICKSEN-FAMILY OFF-KITCHEN ADDITION

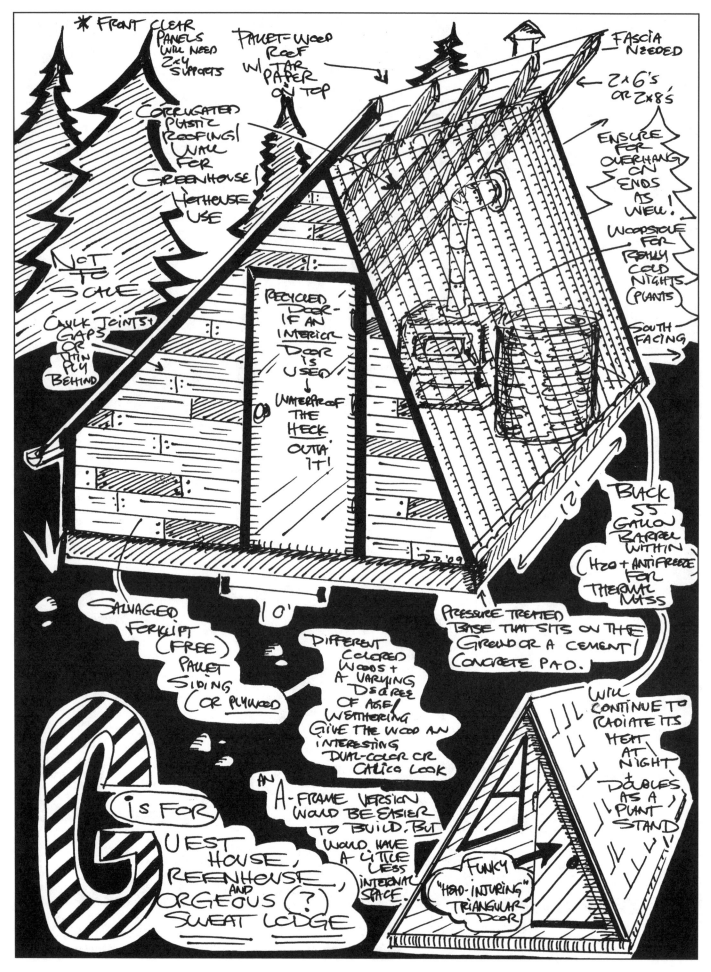

* FRONT CLEAR PANELS WILL NEED 2×4 SUPPORTS

PALLET = WOOD ROOF W/ TAR PAPER ON TOP

FASCIA NEEDED

2×6'S OR 2×8'S

ENSURE FOR OVERHANG ON ENDS AS WELL!

CORRUGATED PLASTIC ROOFING/WALL FOR GREENHOUSE/HOTHOUSE USE

NOT TO SCALE

WOODSTOVE FOR REALLY COLD NIGHTS (PLANTS)

SOUTH FACING

CAULK JOINTS/GAPS OR THIN PLY BEHIND

RECYCLED DOOR - IF AN INTERIOR DOOR IS USED → WATERPROOF THE HECK OUTTA IT!

12'

BLACK 55 GALLON BARREL WITHIN (H2O + ANTIFREEZE) FOR THERMAL MASS

J.B. '09

10'

SALVAGED FORKLIFT (FREE) PALLET SIDING (OR PLYWOOD)

PRESSURE TREATED BASE THAT SITS ON THE GROUND OR A CEMENT/CONCRETE PAD.

DIFFERENT COLORED WOODS + A VARYING DEGREE OF AGE/WEATHERING GIVE THE WOOD AN INTERESTING DUAL-COLOR OR CALICO LOOK

WILL CONTINUE TO RADIATE ITS HEAT AT NIGHT + DOUBLES AS A PLANT STAND

G IS FOR GUEST HOUSE, GREENHOUSE AND GORGEOUS (?) SWEAT LODGE

AN A-FRAME VERSION WOULD BE EASIER TO BUILD, BUT WOULD HAVE A LITTLE LESS INTERNAL SPACE...

FUNKY "HEAD-INJURING" TRIANGULAR DOOR

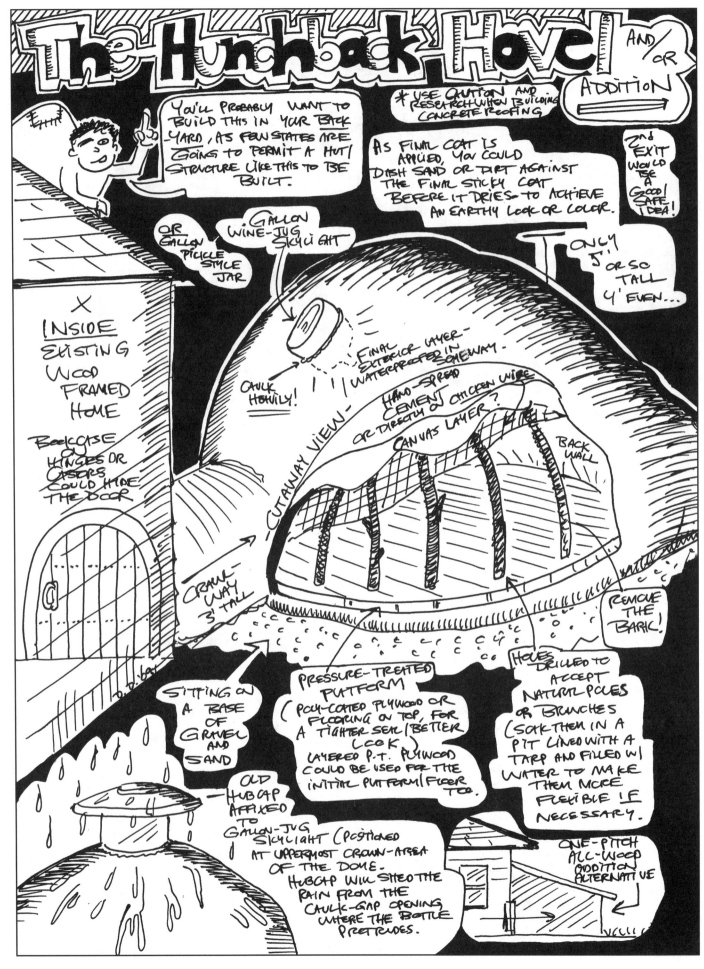

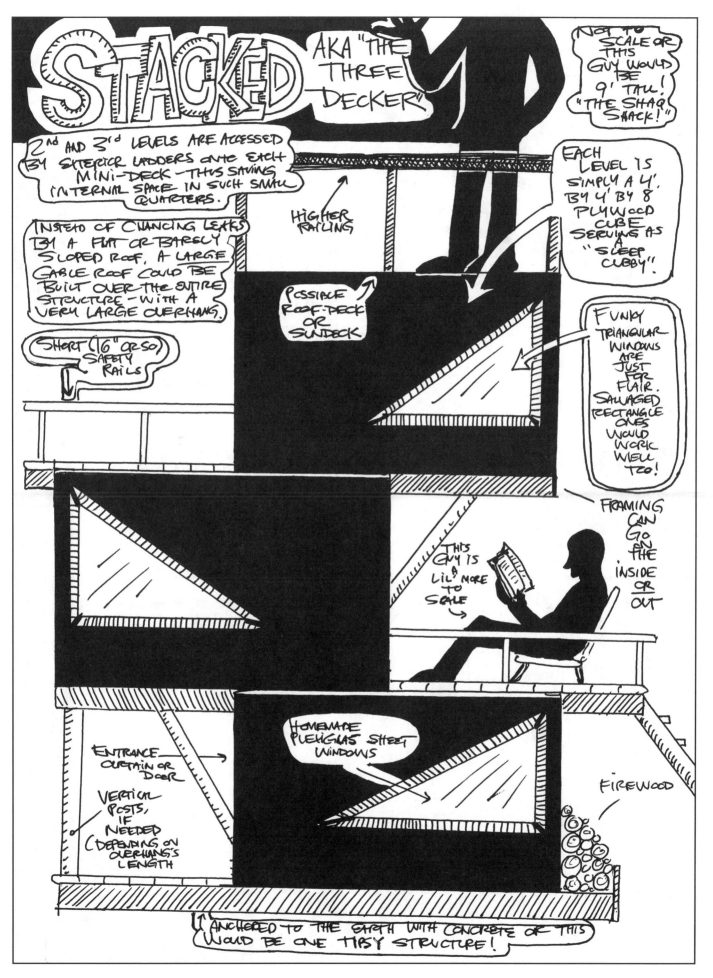

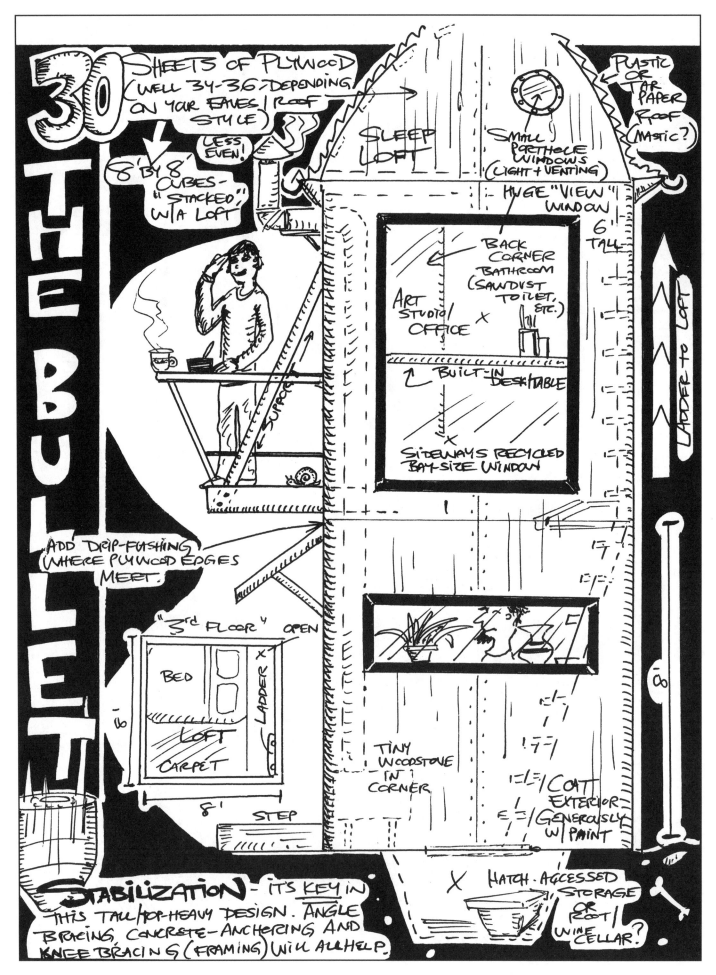

30 THE BULLET

SHEETS OF PLYWOOD (WELL 34-36, DEPENDING ON YOUR EAVES/ROOF STYLE)

LESS EVEN!

8' BY 8' CUBES - "STACKED" W/ A LOFT

PLASTIC OR TAR PAPER ROOF (MASTIC?)

SLEEP LOFT

SMALL PORTHOLE WINDOWS (LIGHT + VENTING)

HUGE "VIEW" WINDOW

6' TALL

ART STUDIO/OFFICE

BACK CORNER BATHROOM (SAWDUST TOILET, ETC.)

BUILT-IN DESK/TABLE

SIDEWAYS RECYCLED BAY-SIZE WINDOW

LADDER TO LOFT

ADD DRIP-FLASHING WHERE PLYWOOD EDGES MEET.

SUPPORT

8'

"3rd FLOOR" OPEN

BED

LADDER

LOFT

CARPET

6'

8'

STEP

TINY WOODSTOVE IN CORNER

COAT EXTERIOR GENEROUSLY W/ PAINT

HATCH-ACCESSED STORAGE OR ROOT / WINE CELLAR?

STABILIZATION - IT'S KEY IN THIS TALL/TOP-HEAVY DESIGN. ANGLE BRACING, CONCRETE-ANCHORING AND KNEE BRACING (FRAMING) WILL ALL HELP.

CLEAR ROOFING FOR LIGHT (ADDITIONAL) INTO "LIVING" (SLEEPING) AREA

METAL ROOF

SLIDING WINDOW DOOR

RAILINGS ON ALL SIDES

SIMPLE STOCK-SIZE 4' X 8' STACKED PLYWOOD CUBES → 3/8" PLY WILL DO → EASIER TO CARRY AND MORE AFFORDABLE $

4'

1'

OUTDOOR "LIVING" AREA W/ CHIMINEA. THIS AREA COULD BE ROOFED AS A PORCH OR LATER FULLY ENCLOSED FOR MORE INTERIOR SPACE.

QUICK DESCENT

2nd TREE TO SUPPORT POINT OF DECK

OPTIONAL WORK/KITCHEN/SLEEP AREA/SPACE

PAD →

FOR SECURITY DON'T PUT WINDOWS IN ON THE FIRST 8' CUBE

4'

4'

STRONG EXTERIOR DOOR- AS THE PURPOSE OF THE "TOWER ENTRANCE" IS SECURITY IN LESS RURAL BUT STILL WOODED AREAS.

TOWER CAN ALSO DOUBLE AS A SMALL LUMBER + TOOL STORAGE AREA (ADD A FEW SMALL BOOKSHELVES, ETC. TOO

DO CEMENT THE BASE POSTS WELL INTO THE GROUND FOR STABILITY

THE High-LIFE

MORE TREE LIVIN'

- UTILIZE THE LADDER-WELL'S WALLS AS A SHOWCASE FOR ART OR EVEN MORE INTERESTINGLY, BAT. ART!

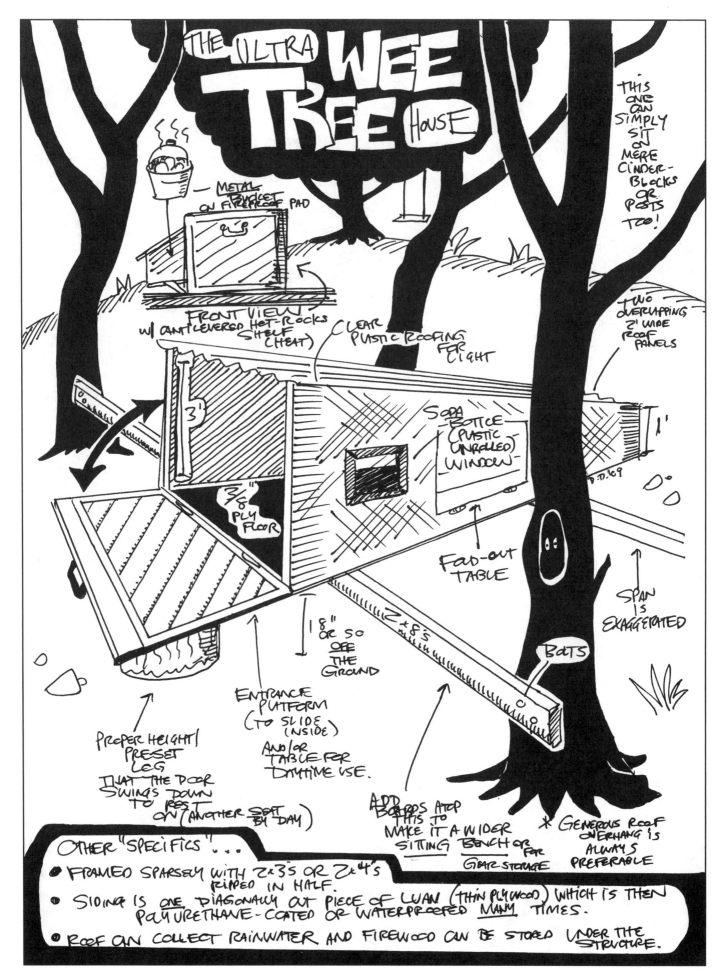

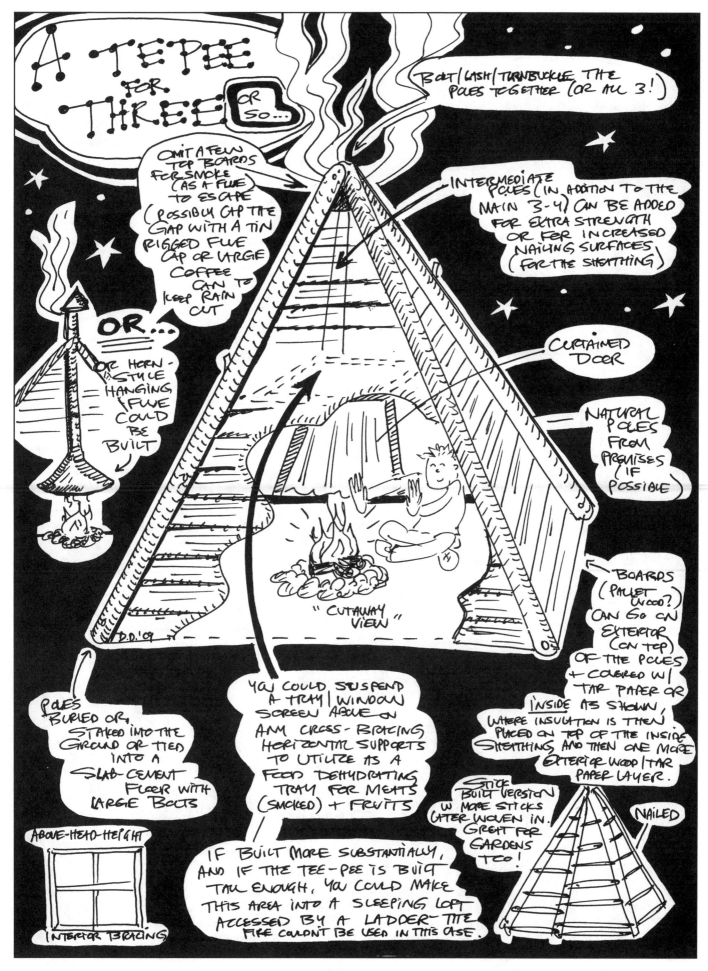

Me in the GottaGiddaWay—a $100 disaster relief shelter I designed *(photo by Bruce Bettis)*

The Hickshaw from "Tiny Yellow House" Episode #1 *(photo by Bruce Bettis)*

Above: The GottaGiddaWay en route to a Cape Cod Eco Fest *(photo by author)*

Below: My backyard scrap wood laboratory/workshop site *(photo by Bruce Bettis)*

Above: My brother
Dustin and me in our
Vermont cabin's loft
(photo by Bruce Bettis)

Right: The interior of
our Vermont cabin
(photo by Bruce Bettis)

Above: The Eco-hab Domehouse *(courtesy of Gemma Roe and Aidan Quinn)*

Below: The E-den in the UK *(created by John Tucker, Glastonbury, UK, www.timbertradesmen.co.uk)*

Above: David and Jeanie Stiles's vacation hideaway *(cabin photo from* Cabins: A Guide to Building Your Own Nature Retreat *by David and Jeanie Stiles)*

Below: The U. B. 30 Treehouse that I built for my brother Dustin's thirtieth birthday as a sleep space in the Vermont woods *(photo by author)*

The "Man Town" mobile storefront on the Massachusetts/New Hampshire border *(photo by author)*

The "Man Town" auto parts store in "closed" mode *(photo by author)*

Left: A micro-house on wheels: George Crawford's Vardo *(photo by George Crawford)*

Below: A Hornby Island caravan *(photo by Michelle Wilson)*

Top: Bill Young's Gypsy Coach *(photo by Bill Young, Lafayette, CO, GypsyCoach.com)*

Above: Nicolette Stewart's caravan in Germany *(photo by Nicolette Stewart)*

Right: A 64-square-foot guesthouse that I designed and built *(photo by Dustin Diedricksen)*

The sleeping quarters (above) and the kitchen (below) of the Whittled Down Caravan built by Tristan Chambers and Libby Reinish *(courtesy of Tristan Chambers and Libby Reinish, whittledown.com)*

Right: Sandy Foster's shabby chic getaway in the Catskills *(photo by Sandy Foster)*

Below left: April Blankenship's cabin in Idaho *(photo by April Blankenship)*

Below right: The home of Graham Strouts in Ireland *(photo by Graham Strouts)*

Andrea Funk's cabin in the woods *(photo by Andrea Funk)*

Above right: A Woods Hole, Massachusetts, nautical shed/cabin

Above left: Martha's Vineyard cottage

Below: Houseboat in Woods Hole, Massachusetts
(all photos by author)

Right: The house of fallen timbers created by David Lottes *(photo by David A. Lottes)*

Below: Amid a birch grove: Tom O'Rourke's cedar-clad cabin *(photo by Kevin Charles Redmon)*

Above right: Lawn Cameron's "Crooked River House" *(photo by Lawn Cameron)*

Below right: A tiny house in Lincoln, New Hampshire *(photo by author)*

Left, top to bottom: Jon Giswold's retreat in northwestern Wisconsin *(photo by Jon Giswold)*

The Sennett bunkhouse in China, Maine *(photo by author)*

Artist/author Cathy Johnson's naturalist's cabin in Missouri *(photo by Cathy Johnson)*

Above left: Dustin and Dawn Diedricksen's Scituate, Massachusetts, home *(photo by Bruce Bettis)*

Above, right: Harrison and Ellie Reynolds's Fern Forest Treehouse, a Vermont B&B *(photo by author)*

Right: A girl hams it up in the GottaGiddaWay. Yes, the window *is* a recycled front-loading washing machine door. *(photo by author)*

Above: A tiny purple cabin at Tilted Acres in Cape Cod *(photo by author)*

Left: Stinky feet not included. Inside the Hickshaw *(photo by author)*

Below: Inside the "Wolfe's Den," a recent guest house in the trees that I built for a client in the Catskills, New York. This was featured on my online TV show "Tiny Yellow House." *(photo by author)*

An art gallery micro cabin built by me and Dustin in Putnam, Connecticut, for a fund-raiser. The front almost "glows" because of the clear Tuftex polycarbonate we used.

My twenty-three-foot-tall Robot Treehouse, deep in the woods of Vermont, was built at a hands-on design workshop I hosted.

The sparse interior of the Robot Treehouse has just enough room on the floor to sleep two people.

Chris Schapdick
of Tiny Industrial
(New Jersey)
and one of his
gorgeous Vardos

A rather inventive conversion (a "skoolie") made from a decommissioned school bus by Blackstone Builders.

Ken Pond of Craft and Sprout Tiny Houses (Connecticut)—a company he co-runs with his wife, Tori.

Facing page and above: A Diedricksen-built treehouse for a police officer's kids in Massachusetts.

Above: Van conversions can be an affordable option for tiny living too (with Daniel and Rachel Messick). *(Banana Van Adventure)*

Right: Just make sure that what you build will suit your stature. Dustin Diedricksen poses in the doorway of a cool little teardrop camper. *(built by Tommy Gerson)*

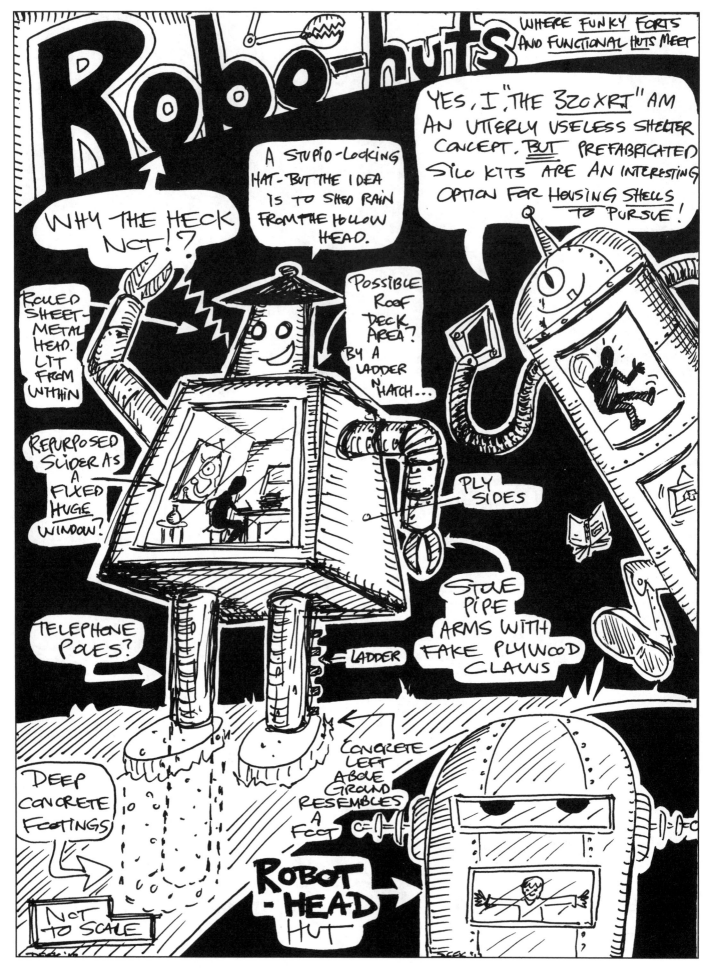

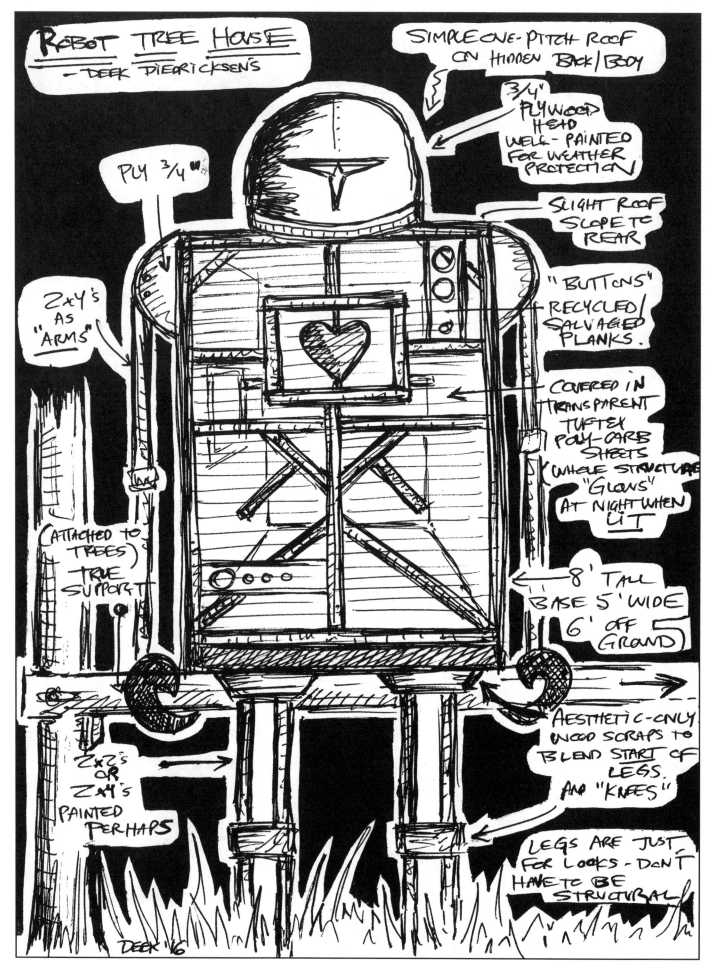

ROBOT TREE HOUSE
— DEEK DIEDRICKSEN'S

SIMPLE ONE-PITCH ROOF ON HIDDEN BACK/BODY

3/4" PLYWOOD HEAD WELL-PAINTED FOR WEATHER PROTECTION

SLIGHT ROOF SLOPE TO REAR

PLY 3/4"

"BUTTONS" RECYCLED/SALVAGED PLANKS.

COVERED IN TRANSPARENT TUFTEX POLY-CARB SHEETS (WHOLE STRUCTURE "GLOWS" AT NIGHT WHEN LIT

2×4's AS "ARMS"

(ATTACHED TO TREES) TRUE SUPPORT!

8' TALL BASE 5' WIDE 6' OFF GROUND

2×2's OR 2×4's PAINTED PERHAPS

AESTHETIC-ONLY WOOD SCRAPS TO BLEND START OF LEGS. AND "KNEES"

LEGS ARE JUST FOR LOOKS - DON'T HAVE TO BE STRUCTURAL

DEEK '16

LETS "FACE IT"

VENT - SO AS NOT TO COOK YOUR GUESTS OR PLANTS...

EACH END IS MADE FROM TWO 3/4" CURVE-CUT PLY SHEETS

8' TALL 8' WIDE

A MORE "TRADITIONAL" SEMI-GOTHIC GREENHOUSE

DOOR (CIRCULAR) NOT SHOWN OR CANVAS BUTTONED FLAP?

POLY CLEAR ROOF W/ MUCH OVER-HANG!

TONGUE N' GROOVE DECK

DECK 2"

HEAVILY PAINTED PLYWOOD TENTACLES

BIRDHOUSE NOSE?

FLOORS

THE EYES? - WATER-COOLER JUGS, PUNCH BOWLS, OR GLOBE FISH TANKS!

SCARY SEA-MONSTER VERSION!

OLD MCDEEKULA HAD A CABIN . . .

An interview excerpt from a feature on the author's own Vermont cabin in *Tiny House Design Magazine,* October 2009 (www.tinyhousedesign.com)

This particular cabin was mainly built by myself, my brother, Dustin, and my wife, Liz. I bought the land up in the Northeast Kingdom of Vermont back in 2000, when I was twenty-three, and built it on weekends here and there. When I could make the four-hour drive we pecked away at the cabin mostly by hand. We used no power tools until we had to cut or rip plywood sheets, which is tiresome and difficult to do with handsaws. A decent portion of the plywood on the structure was donated by a neighbor who worked for a fire salvage/demo company in Vermont. A good many other parts and materials of the cabin were found curbside, such as the exterior door and all but one window.

Most of the other materials were purchased from mom 'n' pop mills up in that neck of the woods where bargaining down a price and haggling are welcome. You gotta love that! I bought the seven true-2-by-8 cedar roof rafters (each initially 15 feet in length), for a mere $40 at one mill. The entire staircase and deck leading to the cabin's entry were built out of bargain-pile/grade 3 cedar amounting to a little over $50. And the cabin overall probably cost in the ballpark of $4,000 to build. Perhaps less.

The cabin's main room is 10 by 10 feet and has a one-pitch roof and a loft above that is accessed by a ship-galley-style ladder (also built from local cedar). There's also a storage and sleeping area beneath the cabin's raised pole structure that we've dubbed "the coffin" for good reason. This tiny room is only 5 by 10 feet in size, with 5 feet of "standing" room, and has one small door and a tiny window. All in all, the existing cabin has approximately 250 square feet of usable space.

We're adding on a new 12-by-16 room, with a 10-foot cathedral ceiling. All the walls for that structure (as I now have two kids and less time to make it up there) were prefabricated in my Massachusetts backyard and simply "Lego'd" together on site.

This method works surprisingly well. All of the prefabricated siding is tongue-and-groove cedar too, and cost me no more than $100 altogether. By building pieces of the structure ahead of time it took only two weekends to get both the walls and the roof assembled. When completed, for use by seven people, the cabin will stand at 450 square feet, give or take.

It's additionally entirely furnished in curbside furniture or castaways from family. The locals have dubbed it the "backwoods skyscraper," as being built on a hill it's 27 feet tall from the lowest point of the hill to the highest peak of the cabin. The metal sheeting wrapped around the base of the cabin, later earning the cabin the nickname "The Diner" by my brother, is the only peaceable way we've been able to keep the porcupines at bay (see: "A 'Terror' to all Tiny Cabins" in chapter 2).

There's an outhouse down the hill, but as for amenities, we have no running water, no electricity, and we try not to allow radios or any "newfangled" gadgets up there when we're camping, as it ruins the solitude. We do plan on driving our own well at some point and with a stream and swamp/pond on the property we feel we have a decent enough chance at hitting water within 20 to 30 feet of the surface.

All in all, it's a 10-acre spread with the nearest neighbor ("Uncle Bob [Bennett]," we've dubbed him, a true seventy-year-old jack of all trades) approximately a quarter of a mile away.

Me, in all my unshaven, stinky "glory," in front of the Vermont cabin Dustin and I built, pre-porcupine assault *(photo by Bruce Bettis)*

DON'T EVEN GET STARTED YET!

Here are a few things you should consider *before* ever starting your build. Yeah, this might seem like a rather bizarre thing to recommend in a book whose sole purpose is to encourage you to just get out there and build, experiment, and try your hand at, well, "anything shelter related," but as I've been doing this for some time, "waiting and planning" is a warning well worth heeding. Even with recycled materials, Dumpster freebies, or the leftovers from old Aunt Edna's shed, you are about to set in motion a project that may take a considerable amount of your personal time. (Actually, it *will*, not *may*.) And who has piles of personal time to just blow these days?

I've long believed that there are a few things you should consider before taking the plunge. The following pieces of advice will undoubtedly save you money, a good deal of labor, and, most important, your sanity. Only the most masochistic individual would want to build something that prematurely collapses or has to be moved to another site. As I've said time and time again, "None of this is rocket science, nor is it quantum physics."

So here is what I often recommend in the "Relax Shacks" hands-on building and design workshops my brother and I have taught for some time now:

1. The Neighbors

That giant fluorescent pyramid fort o' funk that you intend to erect inches away from the property lines might not go over as well with your neighbors as you assume. You don't want to risk not being invited to their Arbor Day oyster bash, so tread lightly and ask politely. And when new neighbors move in, kill them with kindness. Neighbors are less likely to call the authorities if they know and like you. Talk to them before you build and avoid surprising them with that Egyptian-inspired mega fortress.

2. The Law

If you're building anything from this book in a Homeowners Association (HOA) zone, you might as well just pack the idea up, especially if it's out of the ordinary, oddly colored (or multicolored), or, well, "daring" in any way. Most HOAs are pretty darn strict and can contractually limit a lot of what you can do

in a neighborhood. The idea is that it could be tough to sell a home located next to the aforementioned pyramid fort made out of recycled pallet wood and flattened soda cans. While HOAs can be a pain, they do make sense in some cases (I can't believe I'm saying that!), and it's tough to work around them. Remind yourself that not everyone may share your passion for asymmetrical "Relax Shacks." My solution was to buy a home in a non-HOA part of town. Rules schmules . . .

3. The Weather

Build to suit your climate. It's as simple as that. While it may save you money to build with ¼" ply scraps, they are going to bleed heat like it's going out of style. When winter sets in, they may find you in a very "solid" state. Insulation keeps heat in *and* out. Don't be the fonzanoon who builds a south-facing glass getaway near the equator unless you want to lose a lot of water weight quickly. It may seem obvious too, but steps such as strapping down a backyard shed or guest hut will be necessary in tornado- or hurricane-prone areas. Know your zone.

4. Your Orientation

Do you want passive solar gain to help heat your house and save on your bills? How should you orient your structure, fort, reading retreat, or hobby shed if this is the case? What's the arc of the sun in the winter and summer in your region? Do you want to harness the sun's heat or hide from it? Should you have huge windows on one side of your structure or almost none at all? The more windows, the less structural rigidity and privacy you will have. These are all things to consider.

5. Your Site

Because I'm not a descendant of the Rockefellers, I've always been on the lookout for "cheap land," which is often overlooked for its imperfections. With such land you'll need to keep in mind the soil composition (how quickly the site will drain after rain) and the potential for flooding, mudslides, falling rocks, rabid weasel dens, or frequent UFO land-

Take *that*, Wesley Snipes! White, micro-cabin-addicted men *can* jump! *(photo by Bruce Bettis)*

ings. (Okay, beware of *some* of these things.) I once looked at a parcel of land in Vermont and passed on it because, while scenic, it was too close to wetland zones in almost every direction. A parcel of land like this, which seemed too good to be true, wasn't buildable, had terrible and inconvenient access, and would become a haven for mosquitoes. None of this makes for a relaxing summer camp escape.

6. Access

This is the most obvious, yet most overlooked, facet of pre-building. To put it bluntly, can you get to your building site easily? The importance of this varies in relation to the size of what you are building. A cabin you build from smaller pieces is better on rougher terrain or narrow roads or trails. This wouldn't be the same scenario with a structure that requires machinery and sizable lumber deliveries. How close will your pile of lumber be to the construction site? Having to hike in boards will get old quickly if this distance

isn't convenient. These are all things you should be plotting out *before* you build. When looking at land, another thing to keep in mind is the proximity to amenities (gas, supermarkets, restaurants) and to lumberyards. If you have to drive a hundred miles to refresh your hardware and building materials, you better plan each trip carefully. Wood runs won't be last-minute ones when you're building remotely; they will eat up an entire day that otherwise could have been spent on actual construction. Trust me, I've been there.

7. Trees: You versus the "Widow-Maker"

If you're not familiar with this term, a "widow-maker" is a dead or damaged tree that could fall on you and turn your spouse into a widow. If dead and skeletal trees loom over your build site, they could be a risk to you and your tiny shack, house, shed, or shelter. Trees, even thin ones, weigh more than you'd think. Play it safe, know your surroundings, and have these trees removed. Many times your insurance company will even require this before they give you coverage.

Healthy trees, on the other hand, can be beneficial to your site. They provide shade from the sun, protect you from prevailing winds, are an attractive addition to any setting, and can help absorb water on your site. So don't go ballistic and cut down every tree within reach.

8. Prevailing Winds

Know which way the wind blows. Often it will come from a particular direction on your site, and you'll want to build in accordance with that knowledge. While a small napping pod or micro shanty might not be much of a wind catcher, a more sizable shelter won't fare as well if its widest and least aerodynamic side faces the wind. Wind (breezes) can be used to your advantage, however, in hot and humid climates. You'll want to harness cross breezes to keep your structure cooler.

A treehouse is a special case when it comes to wind. The higher you travel up a tree, the more movement your treehouse will receive. If one side of your treehouse is wider than the others, I would choose to position the narrow sides toward your site's prevailing winds.

Building on open hilltops or mountains can be risky too. In fact, a tiny house that was improperly locked down to a true foundation in an open and windy region recently made the news. The result was catastrophic when the structure was lifted and knocked on its side—with the occupant inside! Luckily no one was hurt. Let this be a warning to you: *Know your setting* and *build accordingly*!

9. Do Your Research

Know everything you can before approaching a build. Are there new products on the market that will save you time and money? How should you lay out your structure for proper egress (i.e., the ability to get out of it if there's, say, a fire)? What design will maximize your internal space and be most efficient? How will you heat and insulate your structure? What style roof will you use? Books and instructional online videos can aid you in this quest. It takes time but can also be fun. Don't be that guy or gal who says, "I rushed to build my dream getaway, but if I had the chance to do it over again, I would change the following thirteen thousand things." Some of these projects can exhaust a good deal of your time, money, and patience. Make sure that you "go into battle" armed with as much information and research as possible.

The OTHER "Schtuff"

• Home Built Windows
• Solar Showers
• Rain Barrels

CHAPTER 2

• Bridge Building
• Trash to Treasure Thrift Construction
• Storage Units
• Voodoo Incantations (not really)

• Comics
• Recommended Books
• Websites
• and other assorted time-wasting junk!

A "TERROR" TO ALL TINY CABINS (AKA "STRANGERS IN THE NIGHT")

Heed the warning, ye who plan on building anywhere close to nowhere—dem cute and cuddly porcupines are vicious on the wallet!

During season three at my Vermont cabin, I found myself spending large amounts of time fixing the damage and vandalism that had occurred during my absence. I discovered that my new problem was, in fact, porcupines, after I caught one of the marauders red-handed in the beam of my flashlight late one night. Man, oh, man do these little buggers make one hell of a racket when they're chewin' your newly purchased lumber! I'm talking holes straight through your twenty-dollar sheets of plywood!

Apparently, porcupines, much like deer, love salt. So if you leave one of your axes outdoors over a long period of time, there's a solid chance that you'll return to find it missing its handle. Why? Well, human oils are loaded with salinity, making your tools, or anything you touch, prime eats for these quilled ne'er-do-wells. There have been many reported cases of porcupines eating cabins clear through to the ground over long stretches of time. Now *that's* scary, and good luck explaining it to the insurance company!

I do realize that the porcupines were there first, and I'm basically an intruder on their land, so I did research preventative methods of keeping them away from my little cabin. I tried everything I had

found online. I painted used motor oil on the damaged corners of my cabin (horrible for the environment and groundwater), placed mothballs around the area, and even applied hot sauce to the gnawed-on wood. None of these avenues came even close to working, especially not the hot sauce, as one of its main ingredients is salt. I could almost hear the porcupines laughing at my feeble efforts. I hope that I at least gave those little guys some nasty indigestion.

I finally discovered an incredibly effective solution: Porcupines really don't like lead—especially the kind that's fired toward them at high speeds. I hated doing it, and have a four-count as of this date, but nothing else has worked, with the recent exception of covering the lower extremities of my cabin in sheet metal. It looks completely awful, and has earned the cabin the nickname "The Diner." We've also been forced to encapsulate our outhouse in sheet metal, as the porcupines were especially destructive to it. Its new nickname? "The Bomb Shelter."

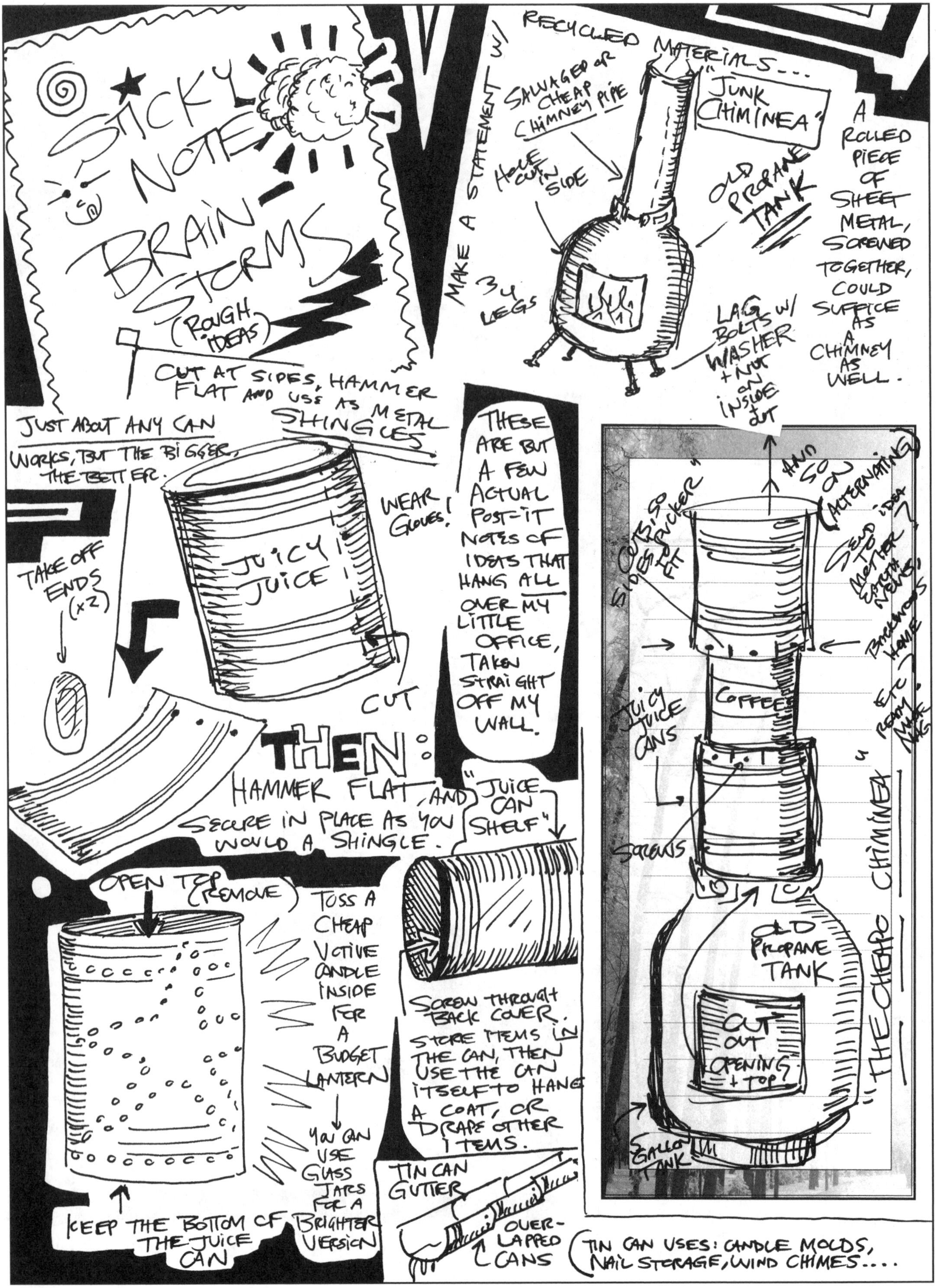

★ STICKY NOTE BRAIN STORMS
(ROUGH IDEAS)

MAKE A STATEMENT W/ RECYCLED MATERIALS...

SALVAGED OR CHEAP CHIMNEY PIPE

HOLE CUT IN SIDE

JUNK CHIMINEA

OLD PROPANE TANK

3/4 LEGS

LAG W/ BOLTS WASHER + NUT ON INSIDE OUT

A ROLLED PIECE OF SHEET METAL, SCREWED TOGETHER, COULD SUFFICE AS A CHIMNEY AS WELL.

CUT AT SIDES, HAMMER FLAT AND USE AS METAL SHINGLES

JUST ABOUT ANY CAN WORKS, BUT THE BIGGER, THE BETTER.

WEAR GLOVES!

JUICY JUICE!

TAKE OFF ENDS (×2)

CUT

THESE ARE BUT A FEW ACTUAL POST-IT NOTES OF IDEAS THAT HANG ALL OVER MY LITTLE OFFICE, TAKEN STRAIGHT OFF MY WALL.

THEN:

HAMMER FLAT, AND SECURE IN PLACE AS YOU WOULD A SHINGLE.

"JUICE CAN SHELF"

OPEN TOP (REMOVE)

TOSS A CHEAP VOTIVE CANDLE INSIDE FOR A BUDGET LANTERN

YOU CAN USE GLASS JARS FOR A BRIGHTER VERSION

KEEP THE BOTTOM OF THE JUICE CAN

SCREW THROUGH BACK COVER. STORE ITEMS IN THE CAN, THEN USE THE CAN ITSELF TO HANG A COAT, OR DRAPE OTHER ITEMS.

TIN CAN GUTTER

OVER-LAPPED CANS

(TIN CAN USES: CANDLE MOLDS, NAIL STORAGE, WIND CHIMES....

CUT OUT SIDES SO THEY STACK & FIT

AND SO ON (ALTERNATING)

SEND IDEA TO MOTHER EARTH NEWS,

BACKWOODS HOME, ETC MAG

JUICY JUICE CANS

COFFEE

SCREWS

"THE CHEAP CHIMNEY"

OLD PROPANE TANK

CUT OUT OPENING + TOP

GALLON TANK

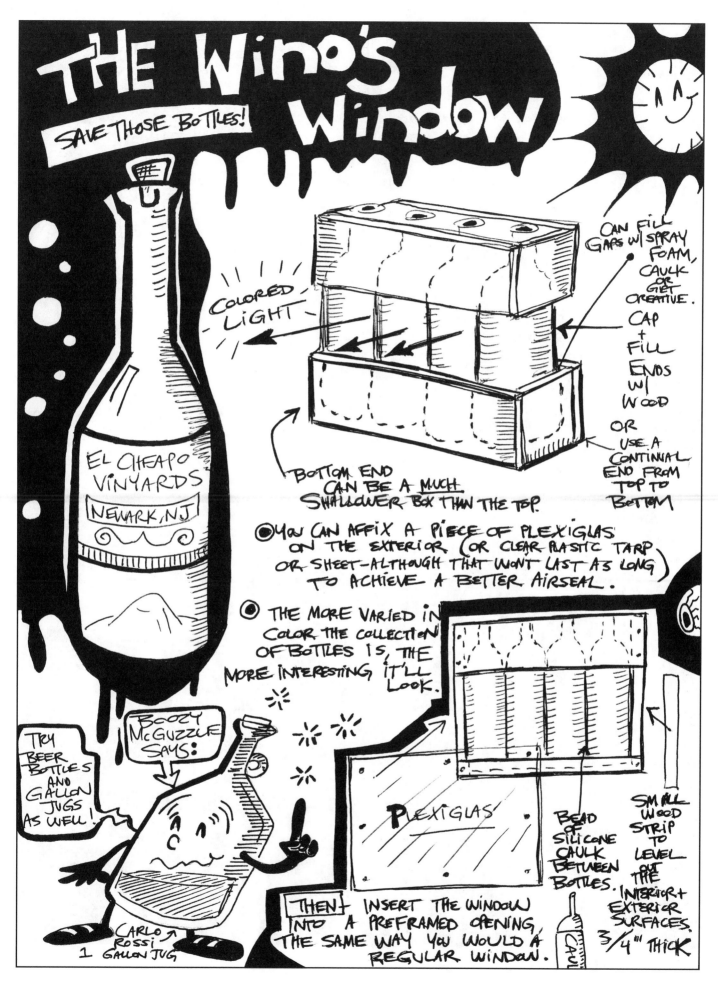

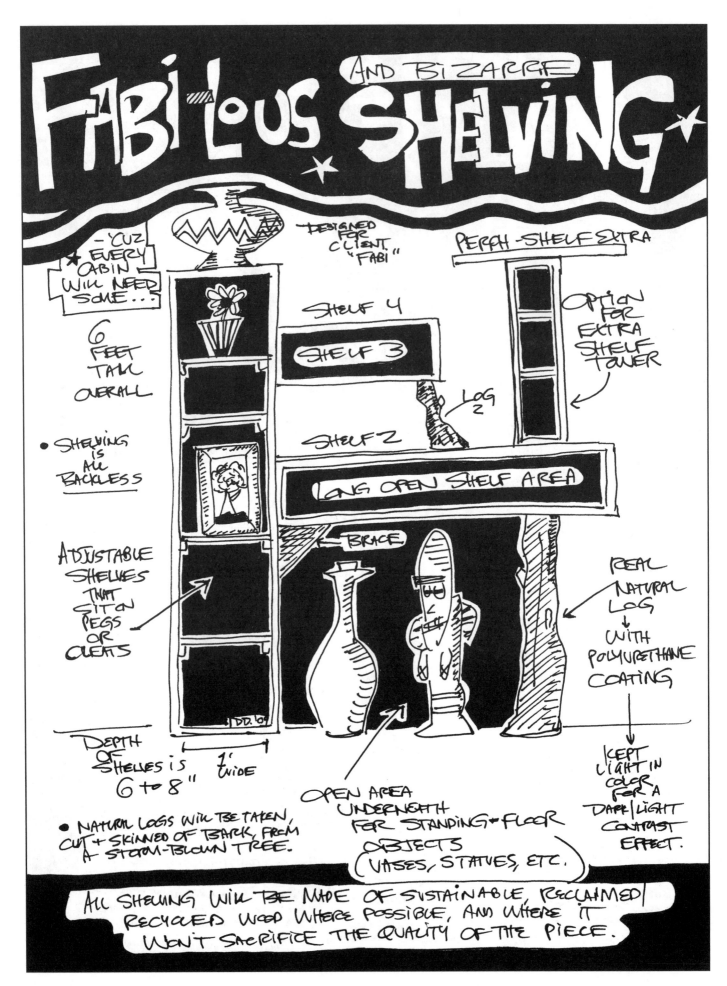

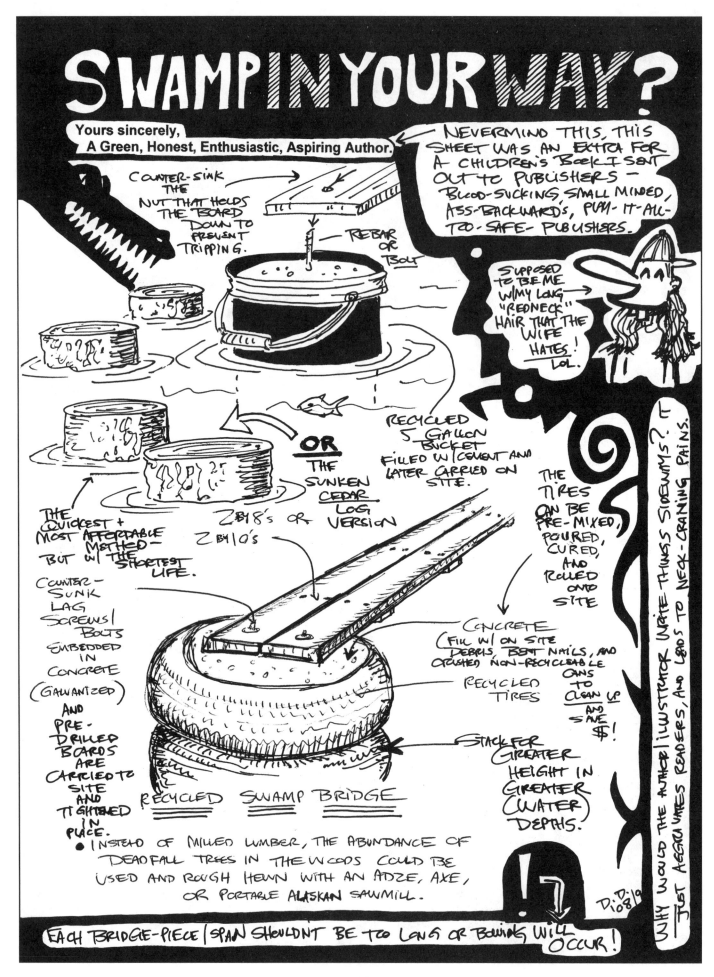

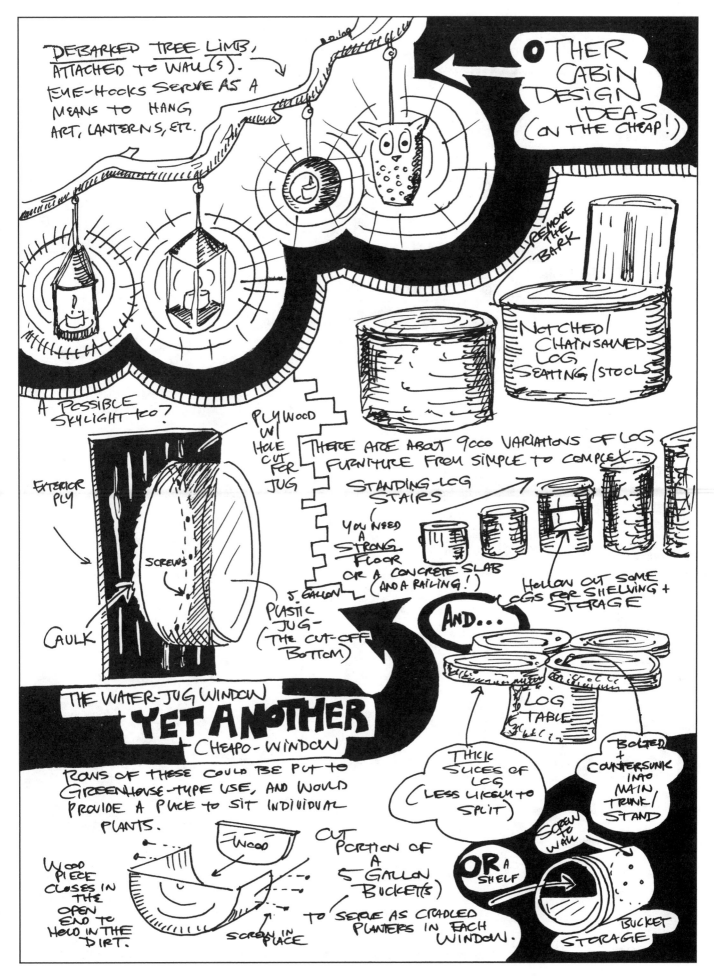

DEBARKED TREE LIMB, ATTACHED TO WALL(S). EYE-HOOKS SERVE AS A MEANS TO HANG ART, LANTERNS, ETC.

OTHER CABIN DESIGN IDEAS (ON THE CHEAP!)

REMOVE THE BARK

NOTCHED/ CHAINSAWED LOG SEATING/STOOLS

A POSSIBLE SKYLIGHT TOO?

PLYWOOD W/ HOLE CUT FOR JUG

THERE ARE ABOUT 9000 VARIATIONS OF LOG FURNITURE FROM SIMPLE TO COMPLEX.

STANDING-LOG STAIRS

YOU NEED A STRONG FLOOR OR A CONCRETE SLAB (AND A RAILING!)

HOLLOW OUT SOME LOGS FOR SHELVING + STORAGE

EXTERIOR PLY

SCREWS

CAULK

5 GALLON PLASTIC JUG (THE CUT-OFF BOTTOM)

AND...

LOG TABLE

THICK SLICES OF LOG (LESS LIKELY TO SPLIT)

BOLTED + COUNTERSUNK INTO MAIN TRUNK/ STAND

THE WATER-JUG WINDOW

YET ANOTHER —CHEAPO-WINDOW

ROWS OF THESE COULD BE PUT TO GREENHOUSE-TYPE USE, AND WOULD PROVIDE A PLACE TO SIT INDIVIDUAL PLANTS.

WOOD PIECE CLOSES IN THE OPEN END TO HOLD IN THE DIRT.

WOOD

CUT PORTION OF A 5 GALLON BUCKET(S)

TO SERVE AS CRADLED PLANTERS IN EACH WINDOW.

SCREW IN PLACE

SCREW TO WALL

OR A SHELF

BUCKET STORAGE

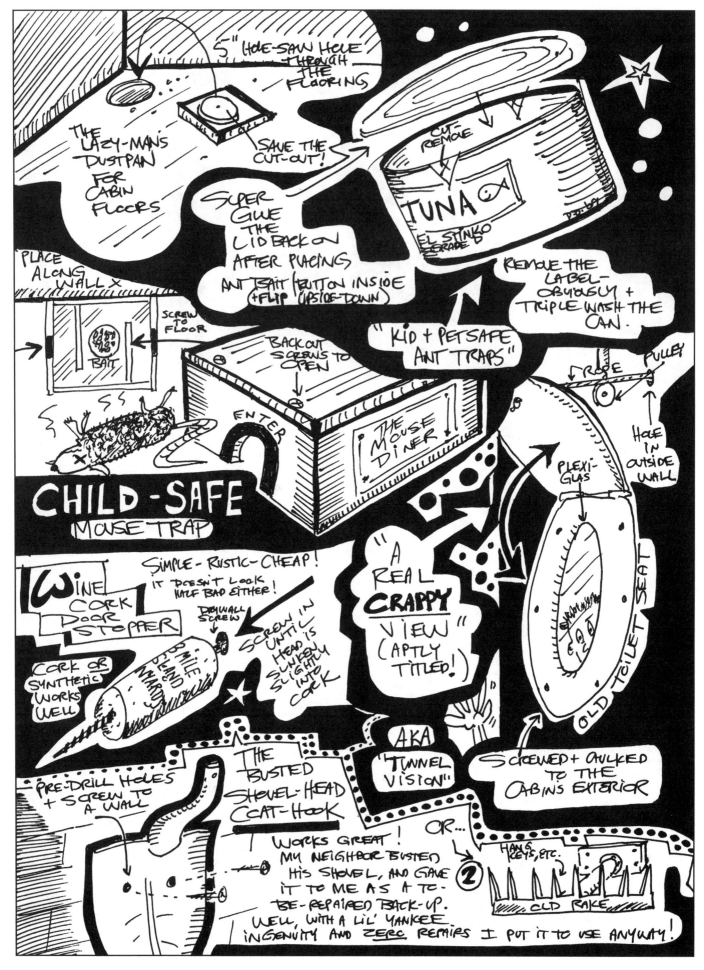

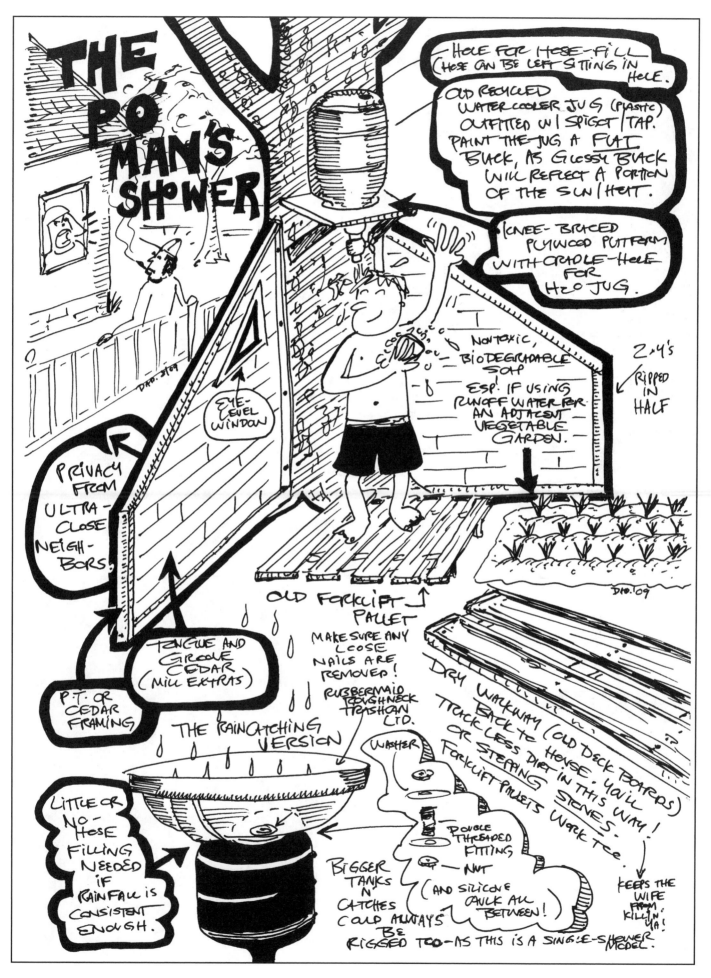

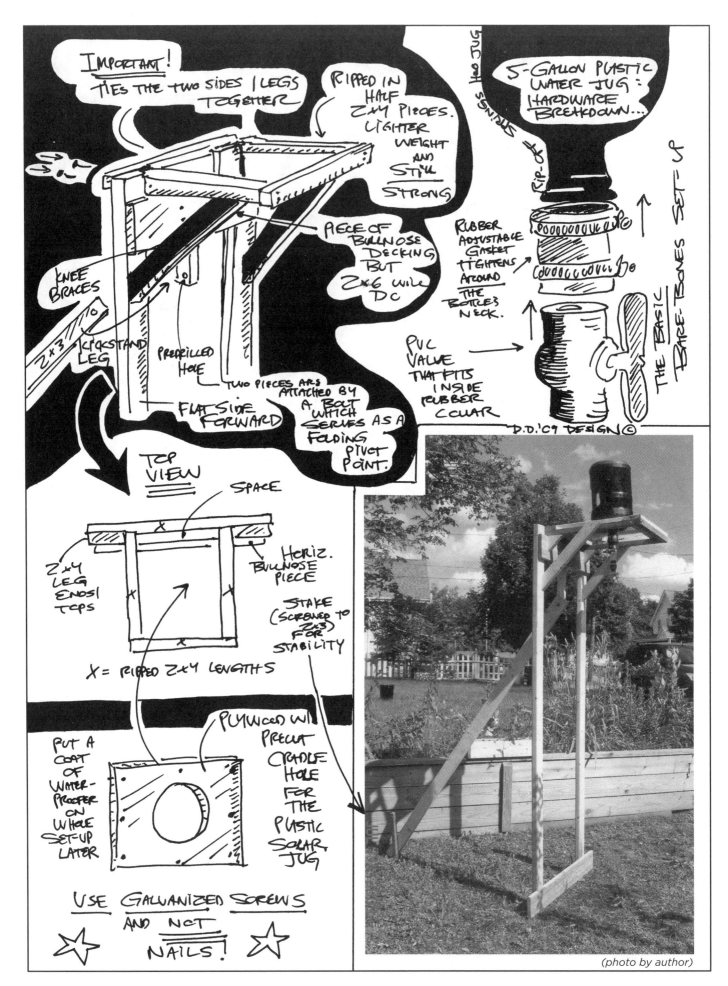

IMPORTANT! TIES THE TWO SIDES / LEGS TOGETHER

RIPPED IN HALF 2×4 PIECES. LIGHTER WEIGHT AND STILL STRONG

5-GALLON PLASTIC WATER JUG: HARDWARE BREAKDOWN...

RIP-OFF STRINGS HOLD JUG

THE BASIC BARE-BONES SET-UP

PIECE OF BULLNOSE DECKING BUT 2×6 WILL DO

RUBBER ADJUSTABLE GASKET TIGHTENS AROUND THE BOTTLE'S NECK.

KNEE BRACES

2×3

KICKSTAND LEG

PREDRILLED HOLE

PVC VALVE THAT FITS INSIDE RUBBER COLLAR

TWO PIECES ARE ATTACHED BY A BOLT WHICH SERVES AS A FOLDING PIVOT POINT.

FLAT SIDE FORWARD

D.D.'09 DESIGN ©

TOP VIEW

SPACE

2×4 LEG ENDS / TOPS

HORIZ. BULLNOSE PIECE

STAKE (SCREWED TO 2×3) FOR STABILITY

X = RIPPED 2×4 LENGTHS

PUT A COAT OF WATER-PROOFER ON WHOLE SET-UP LATER

PLYWOOD W/ PRECUT CRADLE HOLE FOR THE PLASTIC SOLAR JUG

USE GALVANIZED SCREWS AND NOT NAILS!

(photo by author)

82

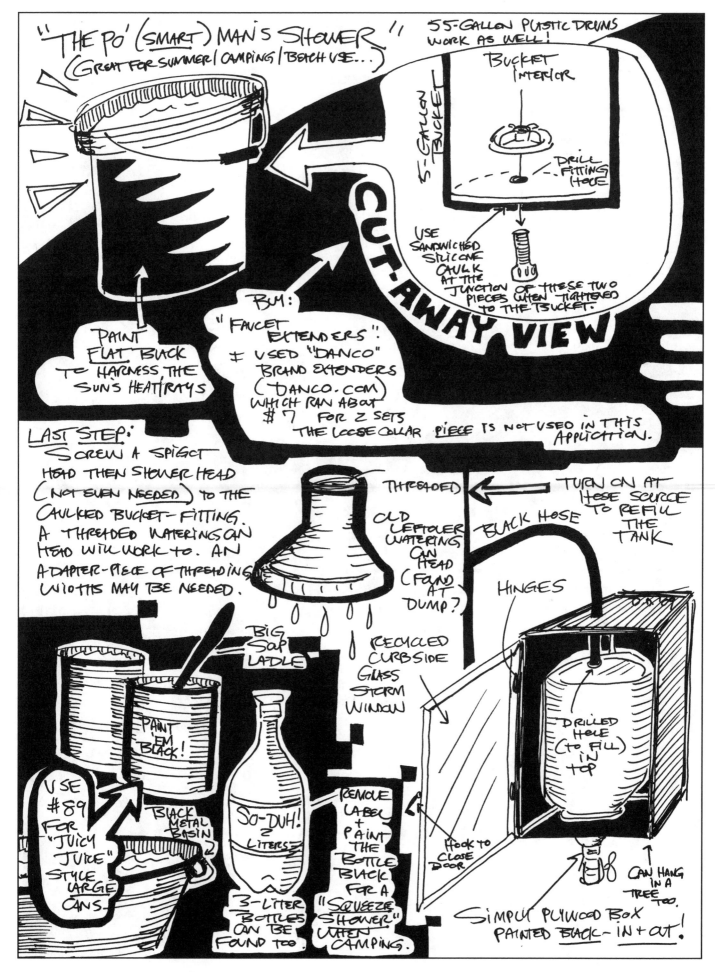

"THE PO' (SMART) MAN'S SHOWER" (GREAT FOR SUMMER/CAMPING/BEACH USE...)

55-GALLON PLASTIC DRUMS WORK AS WELL!

BUCKET INTERIOR

5-GALLON BUCKET

DRILL FITTING HOLE

CUT-AWAY VIEW

USE SANDWICHED SILICONE CAULK AT THE JUNCTION OF THESE TWO PIECES WHEN TIGHTENED TO THE BUCKET.

PAINT FLAT BLACK TO HARNESS THE SUN'S HEAT/RAYS

BUY: "FAUCET EXTENDERS": I USED "DANCO" BRAND EXTENDERS (DANCO.COM) WHICH RAN ABOUT $7 FOR 2 SETS THE LOOSE COLLAR PIECE IS NOT USED IN THIS APPLICATION.

LAST STEP: SCREW A SPIGOT HEAD THEN SHOWER HEAD (NOT EVEN NEEDED) TO THE CAULKED BUCKET-FITTING. A THREADED WATERING CAN HEAD WILL WORK TO. AN ADAPTER-PIECE OF THREADING WIDTHS MAY BE NEEDED.

THREADED

OLD LEFTOVER WATERING CAN HEAD (FOUND AT DUMP?)

TURN ON AT HOSE SOURCE TO REFILL THE TANK

BLACK HOSE

HINGES

BIG SOAP LADLE

RECYCLED CURBSIDE GLASS STORM WINDOW

DRILLED HOLE (TO FILL) IN TOP

PAINT 'EM BLACK!

USE #89 FOR "JUICY JUICE" STYLE LARGE CANS.

BLACK METAL BASIN

SO-DUH! 2 LITERS

3-LITER BOTTLES CAN BE FOUND TOO.

REMOVE LABEL + PAINT THE BOTTLE BLACK FOR A "SQUEEZE SHOWER" WHEN CAMPING.

HOOK TO CLOSE DOOR

CAN HANG IN A TREE TOO.

SIMPLY PLYWOOD BOX PAINTED BLACK-IN + OUT!

SOME OTHER ADDITIONAL, SLOPPIER SKETCHES

HOLE THRU WALL

THE COUNTERWEIGHT WINDOW

CARVED KNOB/PEG

NYLON ROPE

OPENING

20 OZ SODA BOTTLE FILLED W/ SAND

RECYCLED WINDOW SASH + OLD CABINET HINGES

SIMPLE KNOTTED ROPE THROUGH A DRILLED HOLE

SIDE VIEW

GENEROUS ROOF OVERHANG IS NEEDED WITH THIS STYLE WINDOW!

THE LAST OF WINDOWS 'N' WATER

I SWEAR!

D.D.69

r.?? ??

DISTANT HOIST TIE POINT ON GROUND

CAPTURING

SKY-LIGHT

FOR YOUR BACKWOODS SHOWER NEEDS

ONE OF THE TALLER TREES (ABOVE THE SAILS + SHADOWS OF OTHERS) MUST BE USED

(REACHING THE SUN IN WOODED AREAS)

PLACED INSIDE ARE UP TO 4 BLACK-PAINTED 2-LITER SODA BOTTLES FILLED WITH WATER — FOR SHORT "SQUEEZE SHOWERS" IN THE WOODS.

5-GALLON PLASTIC WATER COOLER JUG

HOIST

REG. GALLON BUCKET

SUN

CUT 3 SIDES + PEEL BACK AS A DOOR (WHICH CAN BE SHUT WITH A BUNGEE CORD TO RETAIN HEAT ONCE HOISTED UP

SPIGOT TAP

84

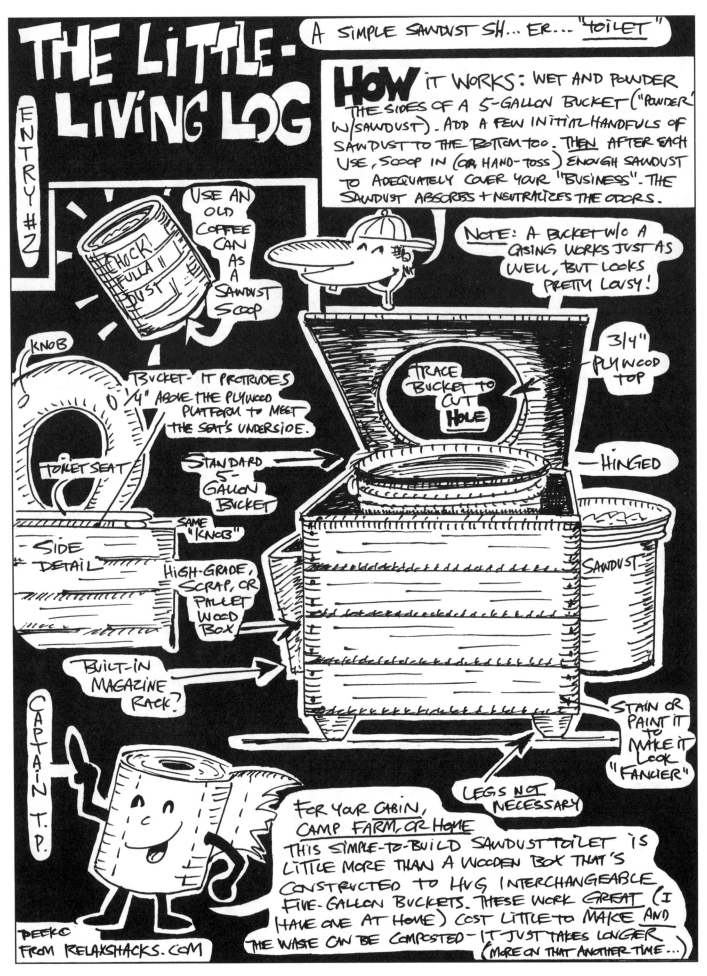

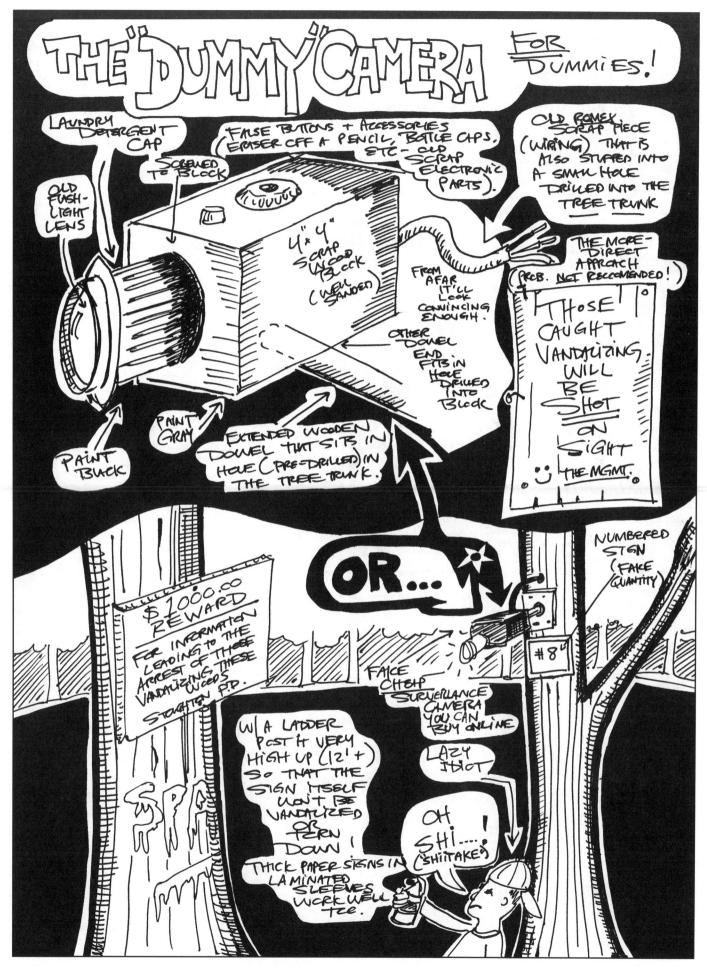

HARVESTING THE RAIN

As tiny-housing and micro-living go hand in hand with leading an affordable life, not to mention treading lightly on the land in many cases, why not set up an easy-to-construct rainbarrel? You can pipe these to flush your toilets, use them for gardening, _and_ utilize them for water needs while off the grid (cement mixing, even drinking perhaps).

ALTERED DOWNSPOUT

HACKSAW THE DOWNSPOUT TO THE APPROPRIATE HEIGHT, THEN RE-USE THE GROUND-LEVEL ELBOW

(HOLE) DRILL A STARTER HOLE, THEN CUT WITH A JIGSAW

METAL ROOFS ARE BEST (CLEANEST). ASPHALT SHINGLES ARE OIL BASED!

THE SPIGOT RIG

3/4" PVC SPIGOT (THREADED-RECEIVING ENDS)

CAULK EACH JUNCTION WHEN ASSEMBLING!

(2) 3/4" LONG GALVANIZED STEEL THREADED PIPE NIPPLES

HOLE CUT BENEATH →

THREADED + CAULKED OVERFLOW (OR TO CONNECT TO 2ND BARREL)

SCRAP-WOOD SCREEN FRAME (SCREWED INTO PLASTIC BARREL)

CLEAN OUT THE BARREL FIRST

55 GALLONS

AND USE "FOOD GRADE" BARRELS ONLY, IF YOU CAN

CAULK THE THREADING (SILICONE CAULK FOR ALL), PUNCH OUT THE CENTER HOLE - AND THREAD IN YOUR ASSEMBLED TAP/HOSE (SPIGOT RIG).

THIS SIDE (WITH THE BUNG HOLES) WILL BE THE BOTTOM!

TO ELEVATE FOR EASE OF USE (HOSE HOOK-UPS, ETC.), PLACE THE BARREL ON A _VERY_ STABLE CINDERBLOCK OR WOOD BASE. A FULL BARREL WEIGHS FAR MORE THAN JOHN GOODMAN (NO DISRESPECT TO THE MAN

DEEK '10 ©

THE SCRAP 'N' CRAP BLURBATHON

The small-cabin, small-living mind-set is often linked with an appreciation of nature and a certain impulse to recycle, the latter of which is frequently rooted in thrift. It is for that reason that I've decided to include this avalanche of recycling ideas, and I believe that it ties into the overall feel of this book. Aside from that, most of these things provide a means of spending less money. Heck, the less you spend (and I'm not pushing you to become cheap, just "smart"—there's a huge difference) the less time you'll waste slaving away at your bone-breaking or mind-numbing job to acquire the adequate funds to survive comfortably. That's the simple basis of most of these varied ideas, and again, pretty much the foundation of this book and its designs.

The following bits of information are thrown at you with no organization whatsoever, but I have included them as I feel they might make for some interesting bathroom reading. Better yet, they just might spark a money-saving, landfill-saving (also less money for trash pickup and disposal) change in you.

Grow your own veggies and save a ton of loot (and eat healthier). Better yet, topple the Burpee empire (honestly, I have nothing against the company) by saving your own seeds. Simply wash them off and soak them in a bowl of water for a short period of time, set them to dry on a paper towel, let them air dry a little more, then bag and label 'em. You'll end up with so many seeds that you can give

Labels in illustration: DETERGENT CAP; YUPPIE WATER BOTTLE; CAP W/ HOLES CRAZY GLUED ON; POOR MAN'S WATERING CAN; PLASTIC 1 GALLON MILK JUG

some to neighbors, friends, enemies, and anyone else you can think of (birds?). Many seeds can be salted, oiled, roasted, and eaten, as well.

If you're short on windowsill space, take to aerial gardening—old plastic juice jugs, mayo containers, and anything of that type can be tweaked into a poor man's "topsy turvy" tomato planter or hanging herb garden. This also keeps plants out of the reach of young children and pests.

It's not revolutionary, but start composting your veggie waste. Your garden and almost-free produce will thank you for it. This also cuts down on trash. For the not-so-squeamish, check out Joseph Jenkins's *The Humanure Handbook*. Just don't tell your friends why your tomatoes are Godzilla sized!

Pizza boxes are great for starting a woodstove, as disposable yard-sale signs, as spray paint goof surfaces, and to use as paper plates (for those wishing to avoid doing dishes whenever possible). Just don't exercise this habit while family or fancy company is over, or you might get 'em talkin' . . . although if you left *this* book on your coffee table, they probably already are! Cardboard boxes, especially the huge fridge ones . . . you know what to do!

"If it's yellow, let it mellow. If it's brown, flush it down" (any other bizarre color, seek medical attention). Most of us learned that little saying in second grade, but if you follow this advice now, you'll save yourself huge amounts of water. The horrible scam that is bottled water deserves a few chapters of discussion, but I won't get myself going on it right here and now. Please recycle those bottles, at the very least.

If you're the carpentry type, save all the sawdust from your projects (as long as it isn't from pressure-treated wood). You can use or even sell it as garden mulch, or make use of it in a composting toilet.

If you're the hoarding and thrifty type (and a big fan of what's in them, as I am), wine bottles can be turned into a vast array of things, such as fancy drinking glasses, chandeliers, and even windows and walls. You can also sell them back to winemakers once you've accumulated a large bunch. Or, if you make wine on your own, you can reuse wine bottles rather than purchasing new ones. You can use old mason jars, jelly jars, and spaghetti-sauce jars as makeshift camp glasses or as storage containers for paint, nuts, bolts, or screws.

Labels in illustration: CUT THE BOX HOLES WITH A SERRATED STEAK KNIFE; FRAGILE; "THE CLASSIC" THE KIDS BOX FORT. GET 'EM STARTED EARLY!; DEEK; "FORT-CYCLING"

91

Large plastic laundry detergent caps (once thoroughly cleaned) make durable plant-seedling pots or great paint containers for kids (as do plastic yogurt cups). Cut detergent containers diagonally and leave the handle in place to make boat bailers, soil scoops, pooper-scoopers, and makeshift toy shovels for sandboxes (the caps can also be used as detail molds for sand castle building).

It's not going to gain you anything monetarily or materially, but I make it a habit to shut off public restroom lights after I leave, especially in joints that aren't often frequented. The same mind-set at home will save you some money.

It is worthwhile to toss old and semi-bent nails into an old coffee can, as you never know when you're going to run out of new/good ones. This has saved my butt many times, and has also kept me from making numerous wasteful trips to the local hardware store. A five-pound box of galvanized nails costs well over $20, and the savings will add up before you know it. Nails that are beyond repair or use can be dumped and recycled at a scrap/metal yard, or can be tossed into concrete footing and slab pours as bulk fill, in which they serve as a binding/rebar-like agent.

Pallet wood is laden with straight nails, and when burned in a woodstove it will reveal its deposit of nails for recycling or even possible future reuse. You can lift these from the ashes with a large magnet. Just shy away from burning galvanized or coated nails at any incredibly high temperature—800 degrees or more—as the fumes generated are not so good to breathe. Pallets can also be made into furniture or compost bin walls, and many other things, and you can find them everywhere!

After you change your car's oil at home (if you do), you can use the motor oil in small doses in your home furnace—just make sure you strain out any particles first. You can find any number of methods online, including straining the oil through a cloth or old shirt, or by wicking via a rope. Also, a few Yankee old-timers still swear by using old motor oil as a stain and siding preservative. I'm not so sure I should recommend this one, as a mere quart of oil can potentially pollute thousands of gallons of groundwater.

Dryer lint makes a great ice-cream topping! No, it doesn't, but you can save it for backpacking and for your woodstove as a great tinder/fire-starting material.

Sticks around the yard should be cleaned up and collected before the winter season if you're a wood burner. Year one, I found out the hard way that you can run out of kindling long before your cord wood runs out. With snow on the ground, good luck finding dry tinder. Without tinder, kindling, and squaw wood (all of which I toss into covered fifty-five-gallon barrels), you're gonna have to suck it up (and cheat) with store bought fire-starters. Old, cruddy, beat-up books (the ones you can't even trade in, donate, or tag sale), make good fire starters as well (using a few pages at a time). Don't burn glossy or colored papers—the "pretty" flames given off won't sit so pretty with your lungs. Also, some comic books and novels from the seventies and beyond were printed with a lead-based ink, so beware—unless a Nero-like

mentality is what you're after! You can also use wine corks as fire starters; just fill an old wide-mouthed bottle with 'em and steep them in rubbing alcohol (as if you were pickling them). Use wintergreen rubbing alcohol for a great scented fire. You can also run a screw through the corks and use them as repurposed door stoppers.

You can affix milk caps to wood sheets using nails/screws/glue to make mosaics, which you can then sell to equally bizarre/artsy/weird people, but the caps also make half-decent washers for screws or bolts. I use them when I make birdhouses from scrap 'n' crap (cans, empty infant formula canisters, carved out buoys, human skulls [kidding!], hollowed-out log chunks, and so on).

Save your old magazines in a pile; those "back issues" always gain some value down the road (like only a year later!). Sell them in bulk on eBay. Oftentimes, the money you make will be more than enough to cover yet another year's subscription. You're also keeping these mags out of the dump.

Egg cartons are great for cheap studio sound-proofing and for seedlings, which can be transferred into the ground, cardboard and all. Old storm window screens can be used as fruit dehydrating racks (if they're aluminum, just lay the fruit on top of a dish towel first). I set apple slices dipped in lemon juice on mine over my woodstove (the lemon juice keeps the apples from oxidizing/browning).

As for walkway deicing, why use those neon blue, chemical-laden, costly sidewalk salts when a nearby beach can provide you with buckets of free sand? It'll be far less harsh on your lawn and your groundwater.

And that's just a short list to get you started.

And last, for the scavenge-savvy, always keep a mini tool set in your car with a multitude of screwdriver bits and sockets, along with a hammer, ropes, bungee cords, and a crowbar, and perhaps a tarp or two (you never know when you'll have to bury a body for the mob. No, seriously, it's to keep your vehicle clean when hauling "finds"). These tools are about all you need in almost any scavenge situation or auto emergency. Call me a "trash picker" if you will, but I've made thousands of dollars—not to mention a couple backwoods cabins—off other people's waste!

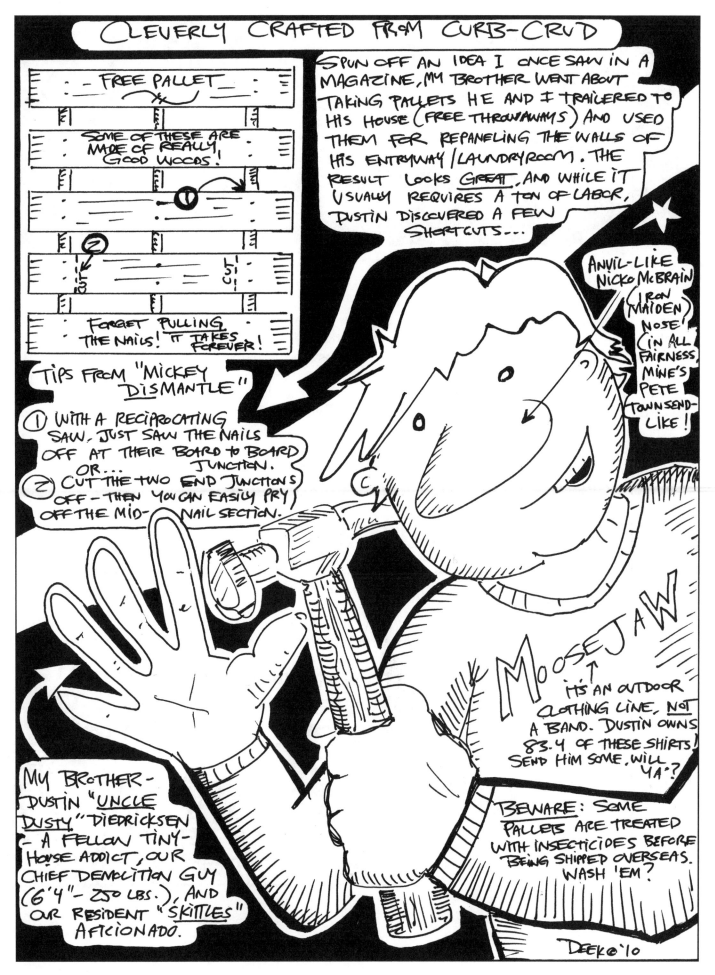

BOUND BY THE POWER OF LAZINESS AND BEER
(Read on, I'll explain, I'll explain...)

BELLY BUST BREW

Thanks for the support and purchase of this book.

Warning: I tend to get longwinded before getting to the point (in this case the book's designs and drawings), so if you wanna just skip over/back to the actual meat of the matter, be my guest; I won't be offended.

While the information I'm about to give you doesn't pertain to this brand-new, expanded edition of the book, I feel it should still be included as I have a minimum word count to meet per my book contract. That is also why this upcoming random sentence has been added to the book: Cantankerous, snorkel-wearing elephants have more fun than left-handed, Kool-Aid-drinking hermit crabs do on Arbor Day. No, no, that's not true at all (the word quota part, I mean. The hermit crab thing is pure fact).

I have ultimately decided that this portion of the book should remain in hopes that it may inspire someone else to follow my example. Perhaps I'll reach a few people with my anti-littering yammering too. All in all, as you'll see, the original self-released small printing and necessary equipment for my book were all made possible by trash. Let me explain . . .

The Beer

If you're one of the ignorant, disrespectful morons who is too lazy to walk three steps out of his or her way to find a trash can or recycling bin, and instead litters on a regular basis, may karma head-butt you with a drunken rhino . . . and thank you! The initial few printings of this book were fully paid for by your

IT BEGS THE QUESTION- "IF YOU'RE A BARTENDER, DO YOU DARE CUT OFF A RHINO?

slothiness. You, Captain Litterpants, are a dirtbag, and I love ya for it!

Without even going gung-ho in my approach, I collected enough beer and soda cans (some even tossed into my own yard) to cover a few early editions of the very book you're holding. Even the $10 comb-binding machine I used to put together those initial shelter-scrawls came from a Boy Scout benefit yard sale at a church—and, yup, littering paid for that too!

It's a simple approach, and one that helped clean up the local woods and hiking trails near my house. I also hope to show you that if no one is willing to publish your work, go it alone! It is possible no matter how broke you might be.

Speaking of which . . . in an economy this depressed, you would think that people would stop littering, if only for self-serving reasons. But nope, a good lot of you have tossed your cash my way instead. It might be only five cents per can, but you're an absolute "Fonzanoon" if you're not redeeming your own cans by now, or at least recycling them.

Yet again, I'm getting wayyyy off track here . . . so here's a comical sketch of a drunken rhino to make up for that. I've pretty much told my tale and made my point. So thank you, good night, tip your waiters, next show's at ten.

Oh, and one more thing.

I'd like to propose that from here on out any imbecile caught littering (like the guy who dumped *several* trash bags full of half-full paint cans and quarts of motor oil in front of a children's school playground earlier this year right next to a beautiful creek in my neighborhood) should have his or her name, address, and mug shot displayed in the local newspaper, the very next day, along with something like the statement I've drafted below:

I, Hubert Snumpy of 58 Wildwood Lane, was caught illegally discarding waste oil near a beautiful wetlands area in town because I care so very little about the environment, the law, other people's neighborhoods, their children, and ultimately you and where you live. That said, as I do cherish my own home (it's why I choose to dump my garbage near yours!), it sure would be a shame if you, the citizens of Stoughton, Massachusetts [or wherever], were to dump all your litter, trash, and hazardous waste on my front lawn, when you happen to be in the area. Sure would be a shame. . . . but I guess there's no way I could prevent it.

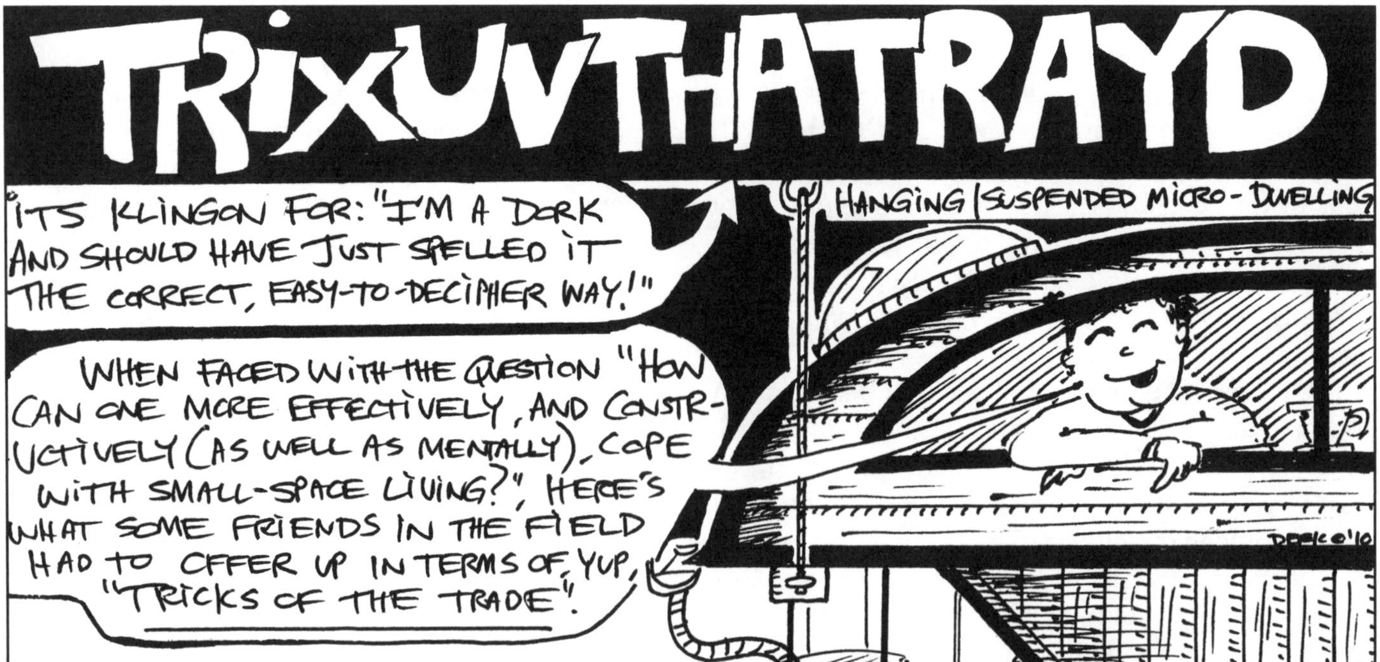

TRIXUVTHATRAYD

ITS KLINGON FOR: "I'M A DORK AND SHOULD HAVE JUST SPELLED IT THE CORRECT, EASY-TO-DECIPHER WAY!"

WHEN FACED WITH THE QUESTION "HOW CAN ONE MORE EFFECTIVELY, AND CONSTRUCTIVELY (AS WELL AS MENTALLY), COPE WITH SMALL-SPACE LIVING?", HERE'S WHAT SOME FRIENDS IN THE FIELD HAD TO OFFER UP IN TERMS OF, YUP, "TRICKS OF THE TRADE".

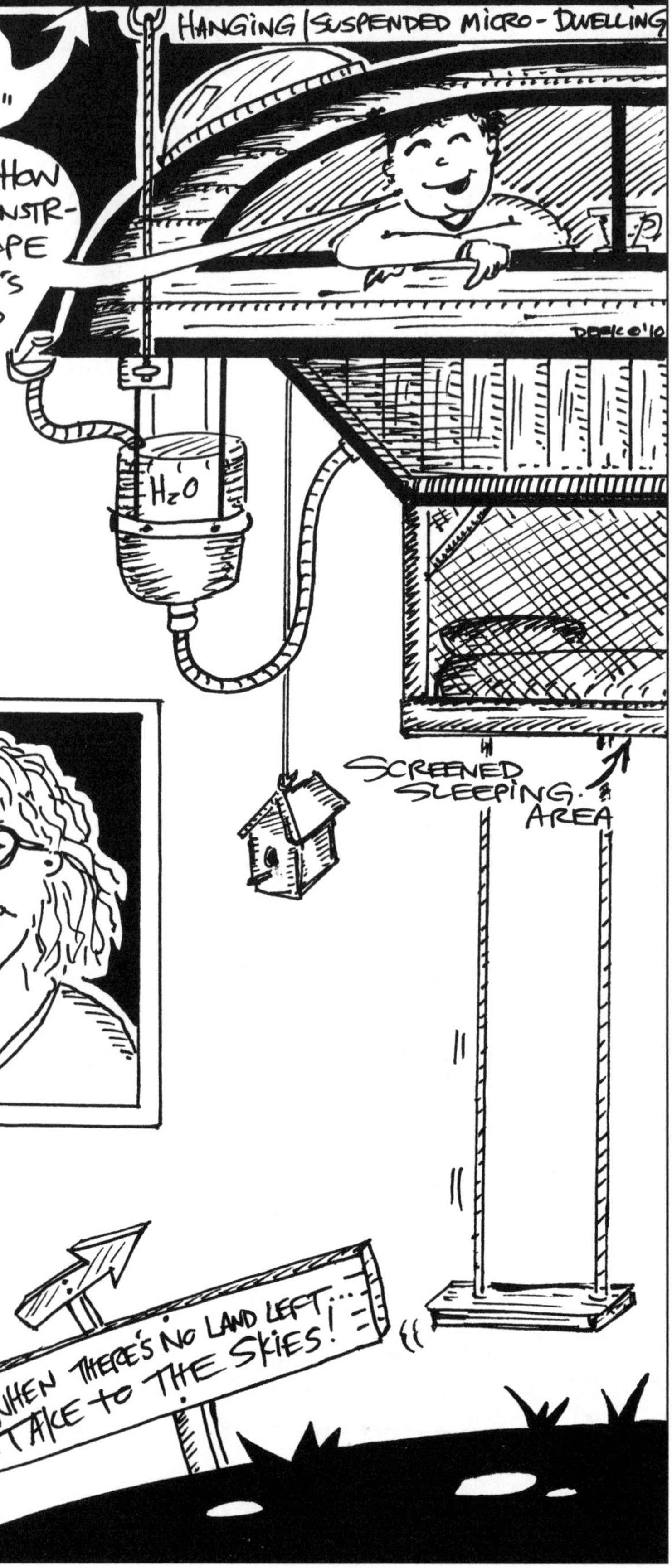

HANGING /SUSPENDED MICRO-DWELLING

H_2O

SCREENED SLEEPING AREA

WHEN THERE'S NO LAND LEFT... TAKE TO THE SKIES!

Mimi Zeiger, author of *Tiny Houses* and *Micro-Green*, writer for *Dwell* and *ReadyMade* magazines

"I love reading novels, but I don't have a lot of space for bookshelves. I take full advantage of the local public library. A friend once advised me to buy vintage furniture because it retains its resale value. I don't have any plans to sell my four white Eames shell chairs, though, since they happen to be my favorite items in the place. They are a bit unwieldy at times, but they add a lot of mid-century design character to the place. My sixties credenza (a Craigslist find) has lots and lots of drawers to store hats, scarves, knitting supplies, and dishtowels."

Michael Tougias, author of
There's a Porcupine in My Outhouse
and *Overboard*

"Extend your living space outdoors. The secret to my enjoyment at my own cabin is its screened porch and an odd-shaped deck. The porch allows me to have that outdoor feel with greenery on three sides, yet be able to sit, write, or eat in comfort. And it's the perfect place to relax when it rains. I do double the living on the deck and porch that I do in my little shack of a cabin. In fact I even sleep on the porch to be closer to nature."

Lloyd Kahn, editor of Shelter Publications and author of *Shelter*, *HomeWork*, and *Builders of the Pacific Coast*

"If I had a piece of land and was just starting, I'd build a central core. Kitchen and bathroom back-to-back, shared plumbing. Woodstove with water heating unit. Solar water heating device on roof. Kitchen opens into garden. Make it small with plans for eventual expansion. Enough so you can cook, eat, sleep, and stay warm."

Kent Griswold, Tinyhouseblog.com

"You have to downsize. Aside from getting rid of things through yard sales, eBay, Craigslist, and Freecycle.org, utilize downscaling technology. If you have tons of photos around, scan them and save them digitally. This also holds true with CDs, as you can easily convert them to MP3s and save them on your laptop. Books use a huge amount of space too, so give them away, use your local library, or go digital."

Cathy Johnson, artist, author of *A Naturalist's Cabin*

"Folding furniture is a great solution to limited floor space, allowing for a great deal of versatility—we have two camp chairs and a vintage folding chair that fits at my desk in my own art shed. The desk has a moveable drafting surface, so it serves double duty. A small oak TV table is just the right height for a laptop, when I need it (the desk is a bit too high). A tabletop can hinge to the wall and fold away when it's not needed. Even my easel is a small folding one—when it's not in use it fits snugly between the desk and the wall. Of course I also planned for outdoor storage—my shed has its own mini-shed attached at one side for construction equipment, barbecue grill, charcoal, tools—you name it. You can store things under the building, too. That's where extra wood is."

Alex Pino, Tinyhousetalk.com

"If I were to give a few tips for living in a small space it would have to be eliminating things you never use, installing shelves, and really liking whoever you live with. Multifunctional furniture like futons, pull out beds, and anything that doubles as storage is a 'must' too. Also, if you're having trouble getting rid of stuff you don't even use just because you're emotionally attached, try taking a picture of it and then giving it away. You'll probably never look at the picture but at least you'll have the comfort of knowing you can."

Gregory Paul Johnson, president of the National Small House Society, director of Resourcesforlife.com

"In terms of the initial downsizing needed, consider taking all your possessions and putting them into a climate-controlled storage facility.

Figure out what you need on a daily/weekly basis, and try living with just those few essential things. If you're ready to take the plunge, move into an efficiency apartment and then begin thinking about what kind of home would meet your needs if you're finding that you can be content in such a small space. Small, planned steps might be the key to a more effective transition."

Deek's Tips

The vacant spaces over household doorways are always ripe and ready for some shelving if you're looking for extra stow space. Also, raise that bed a bit, and you'd be amazed at how many additional things you can store beneath it. Shelving for small items, canned goods, and extraneous doodads (a word I've never used in my life until now) can also be installed under the treads of basement or first-floor stairways, saving space that

otherwise would have been gobbled up by these same items in the kitchen, or elsewhere. Lastly, purchase stackable or foldable guest seats—it's pointless to have many more permanent seats than you'll need, all in anticipation for the potential, and often infrequent, house visitor.

Amanda Kovattana, professional organizer, AmandaKovattana.blogspot.com

"Have nothing that you do not use in your day-to-day life. Rent specialty items, tools, and sports gear

when you need them. Decorate with natural materials and found objects, then return them to nature when you get tired of them. And lastly, pick up mail at a PO box, and write checks the same day at the post office or pay online and you may never have to bring mail home at all.

If you must file something, consider a tiny scanner and keep everything on a laptop."

Duo Dickinson, architect, author of *The House You Build*

"There is one simple rule about designing a small house: It shouldn't feel or function as too small. This means that when you are living in it, looking at it, or simply walking around in it, you shouldn't feel like you gained twenty pounds. The opposite should be true;

you should feel like a burden has been lifted off you. There should be enough light, air, and space around you that you're thrilled to be within the house's four walls. How do you do this?

1. Think about every space and size it to fit the way you actually factually live—shrink the home to *fit* you.
2. Create connections within the interior that allow you to feel the space as a whole, not as a series of little boxes.
3. Make the house permeable. Connect outside to inside to allow for fair weather overflow and 24/7 visual expansion to include the great outdoors."

Jay Shafer, the Tiny Tumbleweed House Company

"For storage, use vertical space wherever possible, and whenever possible. Try to place cabinetry under tables and counter surfaces, while also trying to eliminate staircases altogether—unless you utilize the space beneath them. When designing your living space, it's also key that you minimize transitional spaces and areas such as hallways—they're wasted space. In terms of a home's kitchen, U-shaped counters, where everything is accessible in a single turn, seem to be the best route to take."

Tammy Strobel, Rowdykittens.com and author of *Simply Car Free* and *Smalltopia*

"Living lightly and creatively in a small space can be difficult, especially if you are surrounded by clutter. And when you're just getting started, the idea of decluttering is overwhelming, but necessary. So here's my suggestion: Start small. Take ten minutes out of your day and remove the clutter from one area of your house. Before you know it your junk drawer will be clean, then your closet, and eventually your whole house! Then daily, to maintain, pick up ten things and find places for them."

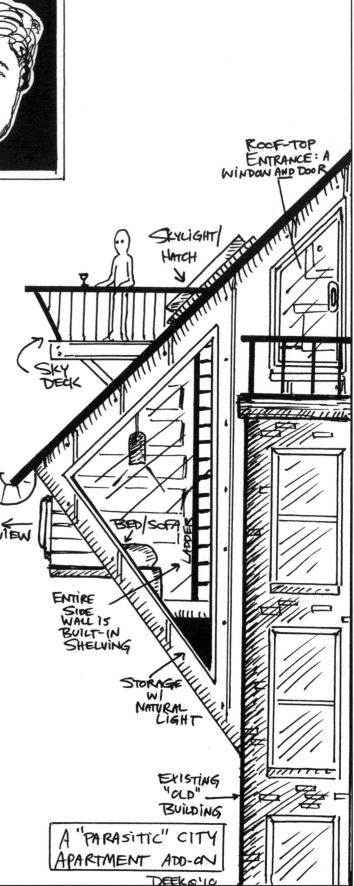

Alex Johnson, author of *Shedworking* and Shedworking.co.uk

"You have to be creative. In a normal office or home there's room for all your stuff to just spread out thoughtlessly, but in a restricted area spots like the backs of doors and under desks can provide vitally useful storage space. And be sure to make yourself comfortable. Don't overlook that aspect of your own well-being. You don't need to lay out millions of dollars, but time and money spent in sourcing or building comfortable chairs, genuinely useful desks, and attractive bookcases will make your space welcoming and conducive to working."

Colin Beavan, *No Impact Man* author, blogger, and documentary filmmaker

"If you're limited on space, there is nothing better than avoiding ownership. Rent and borrow the things you need. Share with friends and family. Use the library for books. Have one toolbox for ten families. One vacuum cleaner. Also, buy services instead of machines. Pay someone else to do your laundry. Then you need no machine. Not only does this all save space but it creates community at the same time."

Dee Williams, tiny house dweller and founder of Portland Alternative Dwellings

"My advice would be to *go small* and *go simple*. Design a house that fits you 340 days out of the year, and plan to improvise the remaining 25 days. I've found this is the case with heating, cooking, showering, hanging out, living . . . it works perfectly and is a dream come true about 90 percent of the time; the remaining 10 percent I am inconvenienced and humbled—just like most of the earth's inhabitants."

P.S. I had hoped to include Dan Price, whose books I love, in this advice section, but after several attempts his publishing company never got back to me. So Dan, if you're out there, make sure you give your promo department a kick in the pants. Maybe next time.

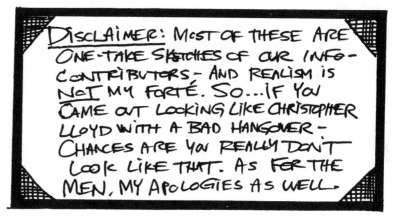

DISCLAIMER: MOST OF THESE ARE ONE-TAKE SKETCHES OF OUR INFO-CONTRIBUTORS — AND REALISM IS NOT MY FORTÉ. SO...IF YOU CAME OUT LOOKING LIKE CHRISTOPHER LLOYD WITH A BAD HANGOVER — CHANCES ARE YOU REALLY DON'T LOOK LIKE THAT. AS FER THE MEN, MY APOLOGIES AS WELL.

PRO TIPS 'N' TINY TRICKS FOR TINY HOUSE BUILDS
WITH SAM AND LYNSI UNDERWOOD FROM
THE SMALL DWELLING COMPANY (WEATHERFORD, TEXAS)

There are a few rules we try to build by. But it all starts with planning and sketching your ideas out. That really is more important than I can stress. I turn to graph paper. My favorite paper is 17" × 22" gridded paper. I like my scale to be one grid square = 3 inches. Our résumé of work shows that you don't need to be a computer genius to put some forethought into what you build. I personally sketch my elevations this way also. I've learned that the more details I draw, no matter how simple a detail it is, the more it helps me grasp the overall design. That said, here are a few rules we abide by:

STAIRCASES

Ask yourself if you can walk up and down your staircase with your eyes closed. Staircases need to be safe. You're going to be ascending them when you are sick, and you will also be walking up and down them at 3 in the morning when you get old and have to pee three times a night. Do you really want steps that rise 12 inches each? You get my point. Don't let safety take a backseat in your design, even if it saves you some square footage space. Keep the rise and runs of your steps comfortable.

AIR MOVEMENT

Air movement is so important in a small space. Chances are that your small dwelling is cooled and heated with a ductless mini. You want that air to move around and find its way in your loft (if you have one), your bathroom, and every corner in your house. It doesn't have to be anything fancy. Something as simple as a desktop fan set in a good location will suffice. If the fan is going to be sitting up high, say on top of an upper cabinet, think about incorporating a plug in your design that's above the cabinet. Have that plug wired to a wall switch and *voilà*! You now have a rotating fan on a switch.

VENTILATION

In your kitchen area you need a vent hood that actually vents outside. I repeat, this vent must vent to the exterior and not into any attic space. Your bathroom area needs an exhaust fan no matter what. The amount of humidity your shower produces will surprise you and later depress you as your tiny home becomes a mold farm and paint begins to peel. Low-ceiling bathrooms (under lofts) need that air exhausted ASAP or that moisture will eventually rise and create other problems in your sleep space.

TOILET 101

I've seen it more than a few times: a toilet stuck in a place that looks like you would have to be in the fetal position to use it. It might be a stinky subject to talk about, but we all require a certain amount of room to do our business. Two things to watch out for: First, make sure you give yourself enough side-to-side room. Second, make sure you give yourself enough "stand up" room. Next time you go to the toilet, take a tape measure with you. See for yourself how much room you need and how much room you can get away with.

My advice to you would be to keep it simple. Stay away from contraptions that require you to use complicated items and maneuvers. Design for the use of simple hinges and things you can fix and correct on your own without specialty tools or help. The fewer moving parts the better is our opinion. Your home is your shelter. It's going to require a ridiculous amount of hard work. You are going to get discouraged and you will have a ton of ups and downs. It's not an easy task to complete on your own, but I can guarantee you that if there are two of you, there isn't anything you can't do.

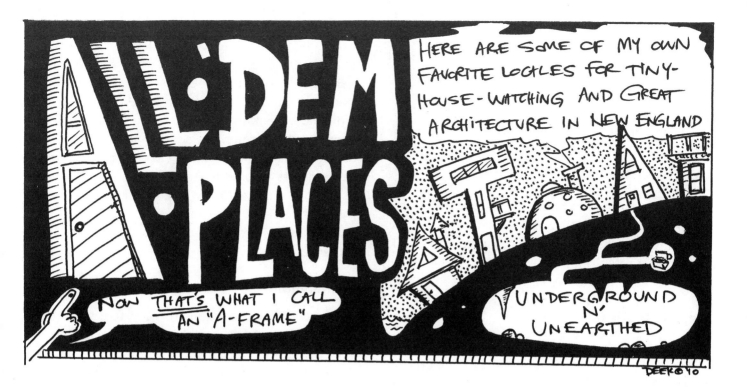

Aside from the list of recommended reading that's included in the Book Nook section of this book, a good many people might be happy to learn of a few off-the-beaten-path locations where they may find tiny and architecturally interesting houses. In the Northeast, I've found that, aside from well-known places like Martha's Vineyard, there exists a hidden wealth of tiny shacks, shelters, cottages, sheds, interesting apartments, and even garage conversions.

Here are seven of my picks from the New England area alone:

Scituate, Massachusetts

It's a beautiful coastal town and its homes are almost on top of one another in some districts. This crowding has prompted some very interesting uses of space and design. Scituate is incredibly scenic, with a walkable and picturesque downtown, and is well worth an excursion on a summer day. This town is also peppered with small, charming, weathered vacation homes from a long-ago era. Its independent bookstore, the Front Street Book Shop, was formerly a small residence itself.

Belgrade, Maine

The Castle Island Camps of Belgrade have been family run and owned for almost one hundred years. These white-sided huts cluster on a miniscule island, connected to the mainland via a bridge. I highly recommend a stay here. Aside from this camp, Belgrade is loaded with hidden lakeside dirt roads that are home to an abundance of cozy looking and unique summer residences.

Madison, Connecticut

When I was growing up, my parents used to take my brother and me on post-breakfast joyrides through the three or four streets that make up a seasonal neighborhood known as Dud's Village in my hometown of Madison, Connecticut. Dud's Village is home to some of the tiniest and most cleverly constructed houses you'll ever see. I really wish someone would make a documentary on this area before the ultra-elite bulldoze it to make way for overpriced condos.

As for Madison itself, it's a gorgeous place to visit, as is the historic town green of its neighboring town, Guilford. Clinton and Westbrook, Connecticut, also offer some great scenic/tiny house viewing—four towns in a row! In Madison, start with Neck Road's side streets. Just don't bother the residents, or tell them I sent ya!

Littleton, New Hampshire

Littleton is an amazing colonial town, still mostly untouched, with a general store boasting "the world's longest candy counter" and a throwback movie theater. Beautiful architecture, friendly locals, and backstreets of quaint, modest-sized homes make Littleton a town I would certainly be willing to move to. It is also near Franconia Notch State Park. Be sure to check out Franconia's Boise Rock if you want to see tiny "housing." Legend has it that Thomas Boise was trapped there in a fierce snowstorm, forced to spend the night under the rock's overhang, protected only by the hide of his freshly killed horse. Yeah, no thanks . . .

Newport, Vermont

On the shores of Lake Mephremagog ("Magog"), this town hides some of the coziest neighborhoods you could ever wish to live in, many with breathtaking views of the lake. Be sure to check out the old-fashioned hardware/department store, the Pick and Shovel—just browsing its aisles is time well spent. Same with the nearby Currier's Market in Glover, Vermont, which is another incredible little town that I'm almost hesitant to tell you about. The store sells everything from guns to soup, and is lined with an enormous array of stuffed (taxidermied) animals. Currier's sits across the street from one of the smallest sit-down restaurants I've even been to (formerly the Busy Bee. It has since changed owners, and is window-service only, but the original structure remains).

China, Maine

Pictorially, I've almost done this one to death, as this book includes a few sketches and photos from this very town. It's where my family and my wife's family vacation every summer. The cabin peepin' down China's various dirt side roads is among the best.

"THE BENDIX CAMP"
CHINA, MAINE

Many of these cabins were hand built by their past owners in the early 1900s and reflect much individual character as a result. The Bendix camp, built by my father-in-law's father, John Bendix, is more "adult fort" than house, making it all the more enjoyable and unique.

Nantucket, Massachusetts

No disrespect to Martha's Vineyard, famous for its tiny gingerbread house community (also well worth the visit!), but I find Nantucket, in terms of house viewing, a little more to my taste and speed. This is especially true on the outskirts of the island, where more of the older, smaller houses still remain. If you're a fan of cedar-shingled homes, you probably won't find more of them in congregation than on this island. Nantucket is pricey, even to visit, but because of some of the incredible small homes it harbors, it should be a must see on your tiny house bucket list.

A REAL HUMBLE, LOW-KEY HOME, BUT THE TAXES (GIVEN THE LOCALE) ARE PROBABLY UNFATHOMABLE!

TINY NANTUCKET HOUSE FROM ACROSS A SALT-WATER LAGOON

I REALLY WANTED TO TRESPASS AND SEE THIS COTTAGE UP CLOSE, BUT DIDN'T...

DEEK '10

Here's a short list of some of the books (and movies and magazines) that inspired *Humble Homes . . . .* In a genre of semi-limited releases, I always appreciate and investigate when other authors refer to a book they especially enjoyed, so here are a few of my own leads for you. Note: These books are listed in alphabetical order by author's last name, but I'd recommend the works of Kahn, Walker, Hubbard, and David and Jeanie Stiles above all others. Some of these may be hard to find, while others are readily available.

Books

Alejandro Bahamon, *Treehouses: Living a Dream*

Ronald Barlow, *The Vanishing American Outhouse*

D. C. Beard (author of *The American Boy's Handy Book* and many others), *Shelters, Shacks, and Shanties*

Cynthia Bix (Sunset Books), *Backyard Cottages*

Jim Broadstreet, *Building with Junk*

Amy Dacyczyn, *The Tightwad Gazette*

P. T. Elliot and E. M. Lowry, *Cracker Ingenuity*

Harlan Hubbard, *Shantyboat* (amazing and incredibly inspiring), *Payne Hollow*, and *Shantyboat on the Bayou*

Joseph Jenkins, *The Humanure Handbook*

Lloyd Kahn, *Shelter, Homework* (with hundreds of beautiful color photos), *Builders of the Pacific Coast*

Ken and Barbara Kern, *The Owner-Built Home*

Dale Mulfinger, *The Getaway Home*

Helen and Scott Nearing, *Living the Good Life*

Dan Price, *Radical Simplicity* (a really cool, offbeat book)

Shay Salomon, *Little House on a Small Planet*

Ruth Slavid, *Micro: Very Small Buildings*

David and Jeanie Stiles, *Cabins* and *Rustic Retreats* (Their books are great, with incredible and fun illustrations. *Rustic Retreats* is among my favorites.)

Sunset Press, *Cabins and Vacation Houses* (difficult to find but loaded with cool tiny homes)

Lester Walker, *Tiny, Tiny Houses* (my first small housing book)

Mimi Zeiger, *Tiny Houses, Micro-Green* (which features my own Vermont cabin)

Non-instructive books and/or books that get me in the building/outdoorsy/adventurous mood (some of the above loosely fall into this category as well)

Edward Abbey, *Desert Solitaire*

T. C. Boyle, *Drop City*

Bill Bryson, *A Walk in the Woods* (and any others; his books are all great)

Vardis Fisher, *Mountain Man*

Dolly Freed, *Possum Living*

Robin Lee Graham, *Dove*

Bernd Heinrich, *A Year in the Maine Woods*

Peter Jenkins, *A Walk Across America*

Henry David Thoreau, *The Maine Woods* (and *Walden,* although it is not one of my favorites)

Michael Tougias, *There's a Porcupine in My Outhouse* and *A Taunton River Journey*

Movies

Garbage Warrior, the Taos, New Mexico/Michael Reynolds story

One Man's Wilderness (I still watch this over and over. There is also a book/journal version, but this is a rare case where the movie is superior.)

Les Stroud's *Survivorman* series on DVD

Magazines

BackHome

Backwoods Home

Boy's Life

Home Power

Make (r.i.p.)

Mother Earth News

Natural Home

The New Pioneer

ReadyMade (r.i.p.)

There are a lot of magazines and books that I do not recommend; however, I do not want to bad-mouth anyone. Be sure to scan book reviews and overviews online before you buy a book in order to see whether it's up your alley.

HAVE HAMMER, WILL TRAVEL

Derek "Deek" Diedricksen was born and raised in Madison, Connecticut, and now resides just outside of Boston, Massachusetts. He is the author of three books in the field of tiny homes and treehouses—this book (duh), *Micro Shelters* (a bestseller), and *Micro Living*—and was a former host and designer for the HGTV series *Tiny House Builders* and *Extreme Small Spaces*. Deek, often with his brother Dustin, has traveled a good portion of the US (all but six states at this point) and beyond, having taught design and hands-on build classes in such far-reaching places as Australia and Hong Kong. He has built, and continues to build, countless "Relax Shacks" for clients, many of them often infused with his original art, and has no plans to stop anytime soon. Deek's work (both art and builds) have

been featured in the *New York Times* (twice), the *Boston Globe*, the *China Times*, and even in the *Sun* tabloid. His creations have also graced the home pages of Yahoo.com, *Design Sponge*, *Apartment Therapy*, the *UK Daily News*, and more. Deek also runs the YouTube channel "RelaxshacksDOTcom," with more than 200,000 subscribers, where he films and edits tiny house tours in his travels. He also fronts the Boston-based metal act Inverter, is a flea market junkie, and collects knickknacks, whimsical robot toys, and other "crud" that drives his wife insane. Additionally, he is the subject/codirector of the documentary project *Box Truck Film*, in which he and filmmaker Alex Eaves construct a full-time dwelling out of a decommissioned moving truck.

(photo by Liz Diedricksen)